introduction to
semiconductor
MARKETING

AL SERVATI AND ANTHONY SIMON

Copyright © 2005 by Al Servati and Anthony Simon. Printed and bound in the United States of America. No part of this book may be reproduced or transmitted in any form or by any means, graphic, electronic or mechanical, including photocopying, recording, taping, or by an information storage or retrieval system with the exception of a reviewer who may quote brief passages in a review to be printed in a magazine or newspaper without the permission in writing from the publisher. For more information, contact Simon Publications, P.O. Box 321, Safety Harbor, FL 34695.

This book contains information gathered from many sources. It is published for general reference and not as a substitute for independent verification by users when circumstances warrant. It is sold with the understanding that neither the authors nor the publisher is engaged in rendering any legal, psychological, or accounting advice. The publisher and authors disclaim any personal liability, either directly or indirectly, for advice or information presented within this book. Although the authors and publishers have used care and diligence in the preparation, and made every effort to ensure the accuracy and completeness of information contained in this book, we assume no responsibility for errors, inaccuracies, omissions, and any inconsistency herein. Any slights of people, places, companies, organizations, are unintentional.

Servati, Al & Simon, Anthony

Introduction to Semiconductor Marketing
Library of Congress Control Number: 2003097802
ISBN: 0-9725189-8-3 (trade paper)
Published by Simon Publications, Safety Harbor, FL.

Book Design by Máximo, Inc., San Diego, CA. www.maximoinc.com

ATTENTION: Educational Institutions, Industry Publications, and Corporations: Quantity discounts on bulk purchases of this book for reselling, educational purposes, subscription incentives, gifts, or fund raising are available. Custom books or book excerpts for premiums and other uses can also be created to fit specific needs. Contact the Sales Dept., P.O. Box 321, Safety Harbor, FL 34695 or email info@simonpublications.com.

ACKNOWLEDGEMENTS

This book could have never been written without the help of our mentors in the industry. We are especially thankful to those who have imparted their knowledge and help in the development of this book. Special thanks to Rich Beyer, Stephen Blake, Peter Kempf, Kaveh Kohani, Bill Peavey.

We would also like to thank those who over many years have helped us develop marketing skills and knowledge including David Andaleon, Bernard Badefort, George (Joe) E. Belch, Lewis Brewster, Steve Cox, Jeff Crosby, Anthony C. D'Augustine, Alain Dubreill, Jeff Ferretta, Greg Fischer, Doug Green, Jeff Hendy, Jean Hu, Jack Jaeger, Amar Kapadia, Chee Kwan, Jim Lumley, Umesh Padval, Matt Rhodes, Bob Sato, Shige Sato, Sumeet Syal, David Tahmassebi, Jeff Teza, Mark Van Zanten, Tim Vehling, Paul Vroomen

We would also like to thank Máximo Escobedo of Máximo, Inc. for the expert design and production of this book.

We thank our wives Hoda and Reiko whose support and hard work made this book possible.

Finally, the authors are dedicating this book to their parents.

TABLE OF CONTENTS

CHAPTER 1: AN OVERVIEW OF MARKETING

CHAPTER 2: SEMICONDUCTOR INDUSTRY ANALYSIS

CHAPTER 3: DEVELOPING A MARKETING STRATEGY

CHAPTER 4: DEVELOPING A ROADMAP

CHAPTER 5: DEVELOPING THE MRD

CHAPTER 6: PROGRAM MANAGEMENT

CHAPTER 7: PRODUCT ROLL OUT AND PROMOTION

CHAPTER 16: ADVICE TO NEW MARKETERS

PREFACE

As we enter the 21st century, the semiconductor industry is finding that the world has become a much different place since the boom years of the late 90's. The world is more competitive, product life cycles are shorter, and product development costs are higher. This era of hyper-competition demands that semiconductor companies place a lot more emphasis on marketing than ever before.

Innovation is important, but useless if it is not directed toward establishing defensible positions and comparative advantages. What this means is that companies must take on a marketing mentality throughout their organizations. The hyper-competitive 21st century requires that everyone in the company be on the same page and fighting for the same mission. Of course, a strong CEO who can impart his vision and instill the motivation for the charge is critical. But, the day-to-day execution of work should be done with the market in mind. Everyone in a semiconductor company should have some idea of how semiconductor markets work.

The hyper-competitiveness of our era requires that engineers, managers, administrators, and marketing people all speak a common language of business and markets. The authors have found that some of the basic tenets of marketing are a real mystery to most people outside of the profession. Most engineers assume marketing is a "black art" practiced by marketers whose business babble should be considered with suspicion. Finance people are less suspicious of marketers, but are often confused why an expensive investment in a new chip sometimes does not lead to consistent profits.

The truth is no one has ever set out to teach most non-marketers about the basic tenets of semiconductor marketing. There has never been any formal training or classes or (until today) books specifically on this subject. Normal business education is too broad. A typical MBA program has little or no information about the nuts and bolts of semiconductor marketing. The authors have set out to change this. The goal of this book is to define and share some time-tested semiconductor marketing concepts. Many of these concepts have never been published in a book but have been passed down from marketing managers to marketing engineers in semiconductor companies over a span of several decades.

This book will give anyone in a semiconductor company the basic ideas, words, and concepts necessary in order to understand how and why marketing people make their decisions. Not only will it help all the players understand how product decisions are made and why, but it will also give non-marketing people the tools necessary to question and challenge marketing decision makers.

Marketing is not an objective activity. Dialog and debate internal to organizations is critical to choose the best path and make the right decisions. However, this dialogue and debate cannot be effective if people in a company do not share the same marketing vocabulary or understand the same basic marketing concepts. This book will help establish a lingua franca, a common language, inside a semiconductor company. It will help companies "get on the same page" and hopefully encourage a team mentality so critical to business success today.

The imperative to get everyone on the same page was not that important in the past. The dot-com era (the last half of the 90's) was an outrageous time for the high technology industry, especially semiconductor companies. It was a period when creating a venture did not require the challenge of marketing a product. In fact, in some cases a product was not even required. The only thing that mattered was stock valuation. At initial public offerings, semiconductor companies with very little history and no proven product market commanded fifteen, twenty, and even a hundred times sales multiple market capitalization.

At that time, sober marketing and decision-making became passé. Semiconductor marketers found themselves making their revenue numbers without doing much to redefine their products or markets. Ever mindful of stock valuations, companies began to use questionable business practices as well as promising the world but delivering little. Concerned as well about meeting earnings goals, companies started "stuffing the channel" by shipping more products than the market needed. These transgressions were thought acceptable because the mantra of the era was that the Internet was changing everything.

The artificial storm has now passed and the glory has been replaced with a hangover that the industry is still trying to recuperate from. While there are some indications of market improvements, stock prices of many companies are now well below their all-time highs. Some stocks have come down from high $100s to below $5. Today a "stock split" is more likely to mean a reverse split ranging as high as twenty to one. Instead of worrying about reaching billion-dollar market valuations, companies now are concerned with survival. In a complete reversal from previous years, Wall Street now values many technology companies at their book, or even cash values.

illustration 1: the dot-com bubble

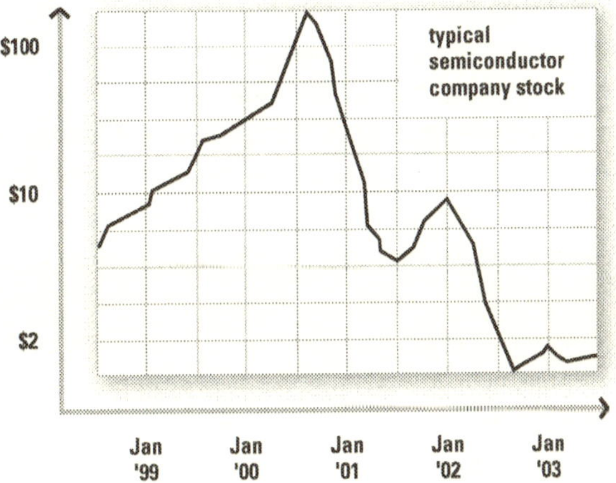

As the semiconductor industry begins to recover, effective marketing is playing a central role in revitalizing the industry. More than ever before, marketers must work effectively to ensure the success of both their product lines and their companies. But they will need help from their colleagues. No longer can marketing toss a requirements document over the wall to engineering and wait for a product to get done. On the flip side, engineering cannot blindly build technology for its own sake and hope that the market will work itself out once the chip is done.

An interest in marketing should not be limited to the marketing department of a company. In the new era of hyper-competition, everyone in a company must adopt a marketing mentality. Each employee must share the common goal of success for their company's products and services. A company crisply executing on its tasks and focused on a set of business goals will always be stronger than a disjointed company with competing objectives.

So who should read this book? All members of a company, no matter if they are in sales, chip design, test and product engineering, system design, applications engineering, finance and legal groups, or manufacturing — all must learn what drives business and understand basic marketing concepts. Everyone must understand how their contributions relate to their company's business goals. This is critical. Just like a chain, every link needs to be as strong as the other for a chain to work successfully.

This book is also useful for people who work with semiconductor companies. For example, suppliers of technology or services need to understand how marketing strategies dictate technology investment in a semiconductor company. Understanding the way semiconductor companies form strategies will help suppliers better align their products and services to the benefit of the industry.

Customers of semiconductor products will also find the material in this book useful. It can help them understand how they play a major role in semiconductor product planning. By appreciating the complexity associated with development and marketing of new products, customers will learn how their input in the early stages of product development could significantly contribute to the success or failure of products.

Semiconductor research analysts and those individual investors who want to know more about the semiconductor industry will benefit as well. The concepts presented in this book will show semiconductor investors how product ideas make it to production and how investors can evaluate the product claims of various companies. Perhaps, more than any other industry, the ability to corner defensible positions is a key success factor in the semiconductor industry. This book will help investors determine what companies are better positioned for success in various semiconductor market segments.

People in technical marketing or managerial roles will also benefit from reviewing the basic tenets discussed in this book. Finally, semiconductor marketers who have recently joined the marketing ranks should read this book.

A semiconductor product's success not only requires having the best-engineered product, but also the ability to market that product effectively. In fact, many semiconductor companies lack any kind of marketing strategy. It is not uncommon for semiconductor companies to develop highly engineered products without checking first to see if there is a real need for such a product. While the foundation of a great semiconductor company is rooted in having innovative, well-engineered products, without a long-lasting strategy shared by all members, a company will have limited growth potential.

HOW TO USE THIS BOOK

There are really only three fundamental activities associated with marketing. These are identifying markets, targeting (choosing) a market, and finding the best way to sell products and services into that market. The semiconductor marketing activities associated with these three fundamentals are explained in detail in this book.

Chapters 1, 2 and 3 take the reader from the basics of market analysis and illustrate how these concepts are applied in practice. The reader is introduced to the key element for marketing success, namely how to establish sustainable competitive advantages in market segments. In order to understand and apply this concept, chapters 1, 2 and 3 provide marketing fundamentals as they are taught in MBA programs. These concepts include defining what constitutes a market, market segmentation analysis, and ways to gauge market segment attractiveness. Chapter 2 provides a survey of the semiconductor market and a review of which market segments hold the greatest promise for growth. Chapter 3 discusses how marketers finalize on a market segment and achieve a sustainable competitive advantage in that segment.

While the first three chapters will help the reader understand how marketers develop product strategy, chapters 4-14 delve into the nuts and bolts of what it takes to bring a semiconductor product to market and ensure the development of healthy and profitable product lines for that market segment. These chapters discuss how a marketer develops and promotes a product roadmap, develops a marketing requirements document, program manages a product development as well as ushers a new product to market.

To help the reader appreciate the work associated with bringing a product to market a good deal of information is provided on roll out strategies, including product pricing and positioning. In addition, marketing considerations regarding products support, forecasting, managing legacy products, and regional markets promotion is also provided.

At the end of the book, chapter 15 discusses the kind of personality and attributes necessary to be successful in the field of semiconductor marketing. Semiconductor companies typically have problems attracting people with the right personalities and capabilities to serve in marketing roles. It is difficult to find people with the requisite technical skills matched with the diplomatic and business skills necessary for the job. Chapter 15 discusses the reasons for this as well as provides an overview of the typical attributes of successful marketers. Finally, chapter 16 provides some advice on keys to success for marketers who are new to the field.

In summary, this book serves as an introduction to semiconductor marketing to those new to these concepts. It is the authors' hope that this information will lead to a greater appreciation for the importance of marketing in the development of successful businesses. By reading this book, non-marketers as well as those new to marketing will learn what and why marketing does what it does. A company's success demands that all members be focused on selling lots of products and beating the competition, this means all members need to at least appreciate the marketing process. This process begins in chapter 1.

an overview of marketing ›

The understanding of semiconductor marketing starts with the study of fundamental marketing concepts. Those with MBAs or similar training may find the material covered in this chapter familiar. These academic concepts are applicable to any market and serve as the intellectual foundation for the practice of classical semiconductor marketing. These concepts also provide a basic framework for marketing decision-making and communications.

The first part of this chapter discusses marketing concepts applied to typical consumer markets. The authors use several consumer market examples to highlight how typical marketing concepts and tools are used. The fact is that the basic approach to marketing is the same regardless of type of market discussed. Later in the chapter and in subsequent chapters we will apply these tools to the technology and specifically the semiconductor market.

UNDERSTANDING THE MARKETING PROCESS

As consumers, we are all exposed to product marketing campaigns every day. These campaigns are often extremely sophisticated and make use of messages and events that, on the surface, may seem random and unrelated. But make no mistake, there is a great deal of thought that goes into product marketing campaigns targeted at manipulating our buying behavior.

However, the marketing campaign is only one aspect of marketing. This may be a surprising notion, especially to those who think that marketing is a combination of sales and advertising. Actually there are three fundamental aspects of marketing: analyzing markets; targeting markets; and finally finding the best way to sell products or services to those markets. Selling and advertising actually come at the end of the marketing process.

As a marketer works his way through the classic steps of analysis, targeting and positioning he is always guided by a golden principle, the holy grail of marketing also called sources of sustainable competitive advantage. The basic meaning of this concept is that a business will always strive to choose a market where for some reason a business has a distinct advantage over any existing or potential competitor. Ironically, players who strive to find market segments where they can be monopolists drive capitalism. We will explain more about this principle as a guiding light throughout this book as it has significant importance in semiconductor marketing.

MARKETING ANALYSIS

In marketing parlance, there are two major aspects to market analysis, identifying and segmenting markets. Let's discuss these concepts in turn.

The process of identifying a market starts with observation. The goal of this observation is to identify unmet customer needs. The ability to recognize unmet needs is a fundamental requirement for successful marketing. Good marketers have the ability to observe and determine if there is a real need for a product or service. We all know that people WANT lots of things, but the real mystery is if the same people who WANT a product or service would NEED it enough to exchange something of value to have the product.

Identifying market needs requires research to understand boundaries that define various product markets. For example, there is a market for serving fast food. There is also a market for serving full-service meals. However, as the failure of Boston Chicken in the 1990s (with its Boston Market restaurant chain) has demonstrated, it is not apparent that there is a market for restaurants that combine elements of the fast-food market with the full-service market.

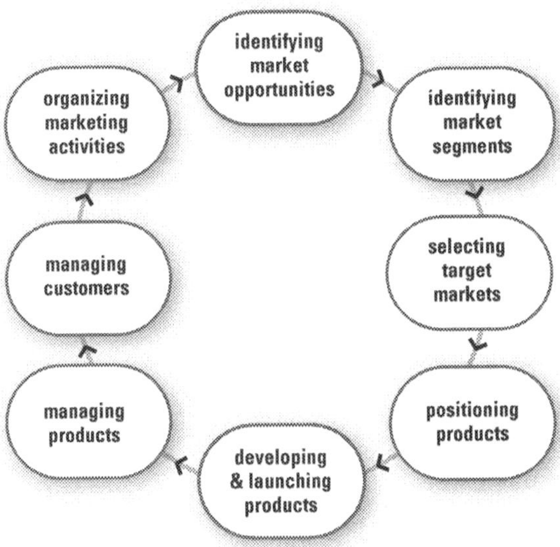

The history of Boston Chicken is a great case study of a company failing to identify a market. This failure resulted in part in the confusion of its customers. Boston Chicken was actually quite successful in the chicken business in the early 1990s. Deciding that serving roast chicken wasn't enough, the company changed the name of its restaurants to Boston Market (confusingly, the holding company retained the name Boston Chicken) and introduced turkey, ham, and meatloaf. In essence, it walked away from a promising market (roast chicken) and chose a totally new market (fast home-cooked meals) without identifying if a market really existed. While rapid expansion, questionable accounting, and poorly managed operations helped in its demise, Boston Chicken's fundamental problem was that customers were not coming to Boston Market stores. In hindsight, Boston Chicken should have stayed in the chicken business and built a business at a more reasonable pace taking smaller steps and identifying new markets before taking an expensive leap of faith.

Consumer needs do not have to be practical or even realistic. "Perception," says the old adage "is reality." The automobile industry spends a fair amount of time observing and learning consumer wants, which they eventually turn into needs. Automobile companies are masters at creating products that people will pay for. Often the industry addresses actual needs (as demonstrated in the development of the convenient minivan, or of fuel-efficient cars). But, the auto marketers also exploit needs that are based on human vanity and selfish desires. The sports car and sports utility vehicle markets are good examples of this. Clearly, we really do not need cars that can take hairpin turns at 90 miles per hour or drive across flooded riverbeds. But we are seduced into believing that we do. We perceive these needs because the car industry has tied the notion of the car to our position, status, and sexual desirability in society.

SEGMENTING A MARKET

After identifying a market, the next step is to further segment the market. Segmenting is necessary in order to draw out detailed market attributes that predict the future of those markets. Of course, it is impossible to consistently predict future trends. However, chances improve for consistent predictions when a market is accurately segmented and well understood.

Market segmentation is a critical step. Without it the subsequent steps of defining your products, estimating Return On Investment (ROI), and identifying sources of sustainable competitive advantage are impossible. Marketing failures and successes are often a function of how a market is segmented.

Market segmentation is an easy concept to understand but very challenging to implement. The reason for this is because segmentation can be a totally subjective exercise. In addition, there are unlimited ways to segment any market. How one company segments a market may be very unique when compared to how everyone else segments a market.

Take the credit card market for example. For years, credit card companies segmented the market by consumer income groups. The higher the income group, the lower the risk of default. No one ever questioned this market segmentation, until one day someone decided to segment the market across race. It was found that Latinos, even though many were poor recent immigrants, as a group paid off their bills on time. Exploiting the market with this new segmentation approach made fortunes for those smart enough to take this approach.

Segmentation is also important when trying to judge the size of a potential market. As a marketer, one is constantly asked to provide market forecasts. Market forecasts are a necessary part of product decision-making and ROI analysis. Without segmentation, market forecasting is impossible; it is simply a wild guess. Segmentation allows marketers to identify attributes of target markets that affect the size and attractiveness of those markets.

Typical market attributes include race, religion, age, occupation, and income of people within a market. (Do not discount the importance of these attributes even to the semiconductor industry. Semiconductor markets targeting games, PCs, and a whole host of consumer applications will be affected by consumer demographics.) In addition, there are also various geographical and political factors that could affect the dynamics of a market. These attributes can be used to segment a market. For example, book publishers know that about 150,000 new titles are published every year in the United States.[1] Publishers also know Americans buy millions of new books every year. While important, these statistics are not sufficient information to make publishing decisions. Therefore, publishers segment their markets when deciding what books to publish. They start looking for more attributes.

For example, they know that people of a certain age group buy history books. They also know that Seattle, New York, and Boston are particularly good history book markets. They have a track record of topics that consumers show particular interest in. They know what kind of cover art attracts history book buyers. The Nielsen rating service is even getting into the book-rating business to supply minute-by-minute sales updates for any book title in any regional market in the United States. Publishers

---

[1]Book publishing statistics are available from RR Bowker LLC at www.bowker.com. Founded in 1872, R.R. Bowker is North America's foremost provider of bibliographic information, and is also the official agency for assigning ISBNs in the United States.

use these attributes to segment their markets in order to maximize their ability to forecast and develop products for each market segment. These attributes must be analyzed, and tools must be created to continuously measure their importance, and to develop responses that allow a company to maintain its position in a market segment.

Entering or creating a new market segment as well as exiting a market segment takes money and time. Major companies with significant resources and long-term strategic goals try to create segments where their products can dominate. Marketers look for segments of various markets that they can dominate and control. Good marketers, in fact, abhor free markets.

TARGETING A MARKET

Once a market is identified and segmented, a marketer must decide which market segment to target. The decision to target a market segment is based on its attractiveness and appropriateness to a company's strategy and market position. Competitive advantage could stem from an inherent strength held by a company. In the technology sector, this is often a patent, or a hold on the best and brightest engineers. Competitive advantage could also stem from creating unbeatable brand equity or reputation in the market. In the most simplistic terms, you target a market where you feel you have the best possibility of success. Maybe this is an untapped market where there are no competitors. Maybe it is a well-known market segment where you have some advantage over the way that business is done.

Selecting a market segment is more complex than simply choosing the largest potential market segment. In fact, the largest markets attract a lot of competition. What looks attractive today can quickly become unattractive, and unless your company is positioned to deal with a lot of competitors, it may be in your company's best interest to avoid that market.

PORTER'S FIVE FORCES

Michael Porter, a respected Harvard University business professor, developed a model called the "five forces model", wherein the attractiveness of a market segment is influenced by:

* Rivalry among the current players that serve that market
* Bargaining power of the suppliers that provide the material to that market
* The bargaining power of buyers
* The threat of new players that plan to, or could, enter the market with similar products
* The threat of new players who plan to enter the market with substitute-products that could significantly change the dynamic of the market

The more powerful these forces are in a market or an industry, the lower the attractiveness of that market. Let's evaluate the domestic airline market segment using Porter's criteria to gauge its attractiveness:

* A large number of full-service and low-cost carriers currently serve the market. This creates a strong rivalry among current airlines.
* Carriers obtain their supply, planes, and fuel from very large companies, often many times the size of the carriers. In addition, there are only two companies that develop planes for major carriers. This means the suppliers to the airlines industry have more bargaining power than each airline.
* The airlines' customers are business and non-business travelers with many choices. This gives airline travelers good bargaining power.
* Low-cost regional carriers continue to enter the industry and are vying to take market share from the big carriers.
* Future high-speed railroads, and increasing popularity of video conferencing among business professionals, could significantly alter the demand for air travel. This provides an alternative to using air travel for conducting business.

No wonder airline margins are razor thin. This environment makes it difficult for airlines to differentiate their service or innovate. Anyone considering entering this market must consider these market factors. By understanding the factors that define the structure and attractiveness of a market, marketers can set the right expectation for their product plan and determine if there is a real chance of competing effectively in that market.

SWOT ANALYSIS

As previously stated, the decision to target a market is also a function of a company's strategy and objectives. For example, we know that Southwest Airlines serves the domestic low-cost air travel market, but it does not have any plan to start a first-class service to Europe or Asia from the United States, even if that market could have attractive growth potential. Southwest prides itself on operating a low-cost airline that focuses on continued profitability without wanting to be known around the world as the largest airline in terms of sales.

A process called SWOT analysis, which stands for Strengths, Weaknesses, Opportunities, and Threats, helps you better target a market. This process requires you to understand the Opportunities and Threats of a given market and match those against the Strengths and Weaknesses of your organization. Chasing market segments that represent large opportunities but do not match your strengths often results in business failure and frustration within the company.

With SWOT analysis, a company chooses a market where its strengths match the opportunities provided by the market. Hence, the company avoids markets with potential threats. SWOT plays an important role in segmenting markets in the semiconductor arena. We will discuss more about SWOT later on in this book.

POSITIONING PRODUCTS

Once a market is segmented a marketer must target one or more specific segments with the right product(s) or service(s). A marketer must decide how to position his product or service versus those of the competition. Products can be defined based on their features, performance, price, and quality. In addition, the amount of service and support provided for a given product can determine how the product is positioned. The last is the quality of the people behind the product and service and the image that is portrayed directly or indirectly by all those involved in marketing a product.

A product's definition, position, and promotion strategy must tie in with competitive advantage. To prosper, companies continuously seek defensible positions for their products. A defensible position is attained because a company successfully exploits sources of sustainable competitive advantage. This concept is particularly important in the semiconductor industry, but it is the goal of any business to be a "lone wolf" or at least "top dog" in any specific market.

Some large high-tech companies such as Microsoft and Intel have had great success creating defensible segments. WINTEL (Refers to the combination of Windows operating system running on Intel microprocessors) dominates the PC industry. Microsoft alone dominates not only the operating system market but has successfully extended its defensible position into application software. For example, Microsoft Office software suite is the predominant office software productivity tool. Microsoft spent well over a decade implementing its strategy of developing an Office productivity market segment where its products dominate. Microsoft continues to dominate this market by continuously raising the bar, making it nearly impossible for a new entrant to successfully raid this segment (at least at the time of this writing).

In the consumer market, Southwest Airlines positioned itself as the low-cost air carrier in the United States with an extensive network and a staff that is known to be friendly. The airline's marketing campaign in the last two decades has been targeted at making travelers think of Southwest first for domestic travel. The airline has also created a simple pricing policy for its fares that gives customers the image of an airline whose pricing is fair and a good value. Southwest has been very successful at dominating this position and fending off competition.

Many companies are less adept at product positioning. A company with no product positioning or well-defined target market will find it difficult to establish a defensible position against current and new competitors.

In the domestic airline market, America West Airlines can be seen as an example of a company that lacks product positioning. It is not clear how the company wants to be perceived. It seems like its strategy is similar to Boston Chicken's: full service yet low cost. But it cannot compete with Southwest on the low-cost side, and it is too small to compete with American Airlines on the full-service side. So the company continues to struggle to find its niche. On the other hand, it is too early to say whether Jet Blue, the new low-cost carrier, can successfully compete with Southwest; however, they seem to be onto a market need that Southwest has not responded to. In a sense, Jet Blue is trying to marginally improve the low-cost model by providing assigned seats and perks like on-board entertainment.

THE FOUR Ps OF MARKETING

There are some basic tools available to a marketer for positioning a product. These tools are often referred to as the four "Ps" of marketing: Product, Price, Place, and Promotion.

By "product" we refer to the form, fit, and function of a product that distinguishes it from the competition or similarly positioned products. For example, Apple Computer spends significant research and development dollars to design products that are appealing to its customers. Apple's industrial designs are some of the best in the world. Sony Corporation also spends significant time on the design of its products. Dell Computer Corporation, in contrast, does not stress design; Dell differentiates its product in terms of ease and flexibility of buying the features a customer needs. Dell has also distinguished its products by providing outstanding after-sales service and support.

"Price" is another significant part of the marketing mix for most products. Airlines depend on their pricing policy to fill airplane seats. Southwest Airlines often offers significant discounts for fares that are purchased months in advance of the travel date. But, lower prices do not always guarantee success for a product or a service. Consumers sometimes view lower prices with skepticism. Sometimes consumers think a low price is a sign of low quality, poor service, or even a potential swindle.

For example, some business travelers refuse to travel on Southwest Airlines because they associate the company's low-cost fares with poor service.

"Place" refers to the distribution channel of a product. Generally, companies sell through retail, to businesses, or directly to consumers. There are also indirect channels, such as a mobile phone operator leasing products to consumers in consumer electronic stores. Most consumers will pay a premium for convenience. That is the reason why many people still buy from their local supermarket or convenience store rather than going to wholesalers, even though they can save money by driving a little further to a wholesaler's location. In the semiconductor industry, most products are sold either directly or through distributors to large and small companies that design and/or manufacture electronic products. Hardly any semiconductor products are sold directly to consumers. Most established semiconductor companies have an in-house sales force that allows their customers to buy directly from the company.

Last on the list is "promotion". Depending on an attribute of a market or a product, often a promotion budget consumes a major portion of the total funds needed to make the product a success. For example, movie studios spend major time and money in promoting their movies, knowing that there is only a small window of opportunity for the success of their products. The same is true for some high-tech products. A good example is when Microsoft introduces a new version of its Windows Operating System (OS). Often the advertising campaign surrounding the launch of a new OS resembles that of a blockbuster movie.

However, most semiconductor companies do not spend large sums of money promoting their products to the masses, given the small number of companies that buy their semiconductor products directly. Intel is certainly one exception to that rule. Intel spends a large amount of its marketing resources to communicate its product features and functionality to the general consumer. Intel cannot afford to let customers decide when they need a faster processor. Because Intel can put more features in a given chip every year, the company has to create demand for those features. Also, knowing that others could create a similar product to its microprocessor line, Intel has spent millions of dollars for the "Intel Inside®"

campaign to make the end customers believe that its products are better than those of its competitors.

Product features, place of distribution, pricing of products, and product promotions are all important factors in image development. How a company or product is perceived in the minds of customers is a major determinant of success in the business world. For example, the airline industry is a difficult place to innovate or differentiate. Yet, there are huge differences of opinion about various airlines. Southwest Airlines, for instance, has a folksy reputation for providing value on domestic flights. Southwest has invested much to achieve its operational excellence and this investment is paying off in terms of a great image in the mind of consumers. Most people believe that Southwest provides low prices and good value. This reputation is one of Southwest's most valuable assets in the business world and a great source of sustainable competitive advantage for the airline.

Without knowing it, almost all of us participate in the process of identifying, segmenting, and positioning for the job market. In high school and college, we develop skills that we believe the job market will find useful upon our graduation. After high school or college, we target markets with the goal of maximizing our skills and talents. The process of marketing one's abilities is similar to marketing products. In both cases we create a "product" that we believe others need and want — a product others will pay for.

Semiconductor companies often overlook these basic tenets of marketing. Too often, they become impressed with their technology and capabilities. This pride leads them to build products for the sake of the technology, believing "if they build it, the customers will come." Many semiconductor products fail because companies overlook the critical step of identifying, segmenting, and positioning of their target market.

Technology for technology's sake is important in the field of research and development (R&D). Some semiconductor companies, governments, and universities can afford pure research. Research is critical to drive creation of entirely new products. We would never have the computer, the transistor, and a whole lot of other inventions had it not been for high-level R&D. However, if a company plans to make money and survive, products must be built for today's and tomorrow's markets, that meet customer needs.

IDENTIFYING UNMET NEEDS

All the key aspects of general marketing previously discussed directly apply to the semiconductor field. Hence, semiconductor companies must identify markets and segment them carefully. Once a market is targeted based on the principles of competitive advantage, it is necessary to define and position a killer product that best fits the market requirements.

illustration 3: identifying customer needs

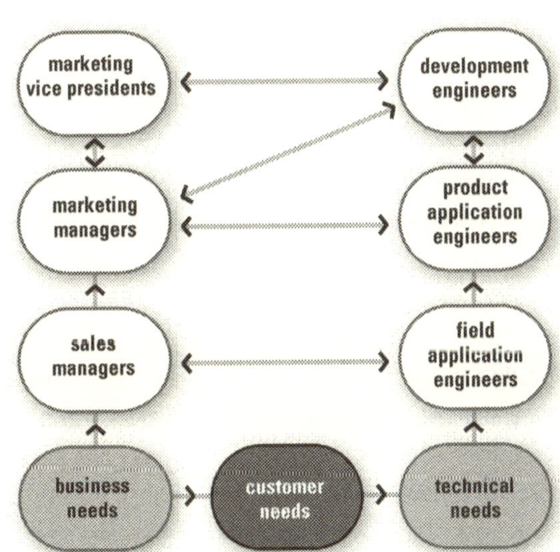

Semiconductor companies should find attractive markets before committing money on research and development. It's easy to spend lots of money, but difficult to earn the money back through product sales. In the early days of semiconductors, it was easier for semiconductor companies to concentrate on technology rather than markets. After all, in the old days when circuits were dominated by discrete components it was easy to find ways to integrate functionality in Integrated Circuit (ICs). But those days are long gone. Business and marketing considerations must take command of technology decisions. As we state in the preface of this book, semiconductor companies must develop a marketing discipline. Everyone working for semiconductor companies must adopt a marketing mentality. Companies should teach marketing concepts throughout their organizations and focus on execution and customer service. "Customer needs" must be communicated at every level in the organization.

RACE TO INTEGRATE FEATURES

There are some differences between general consumer marketing and semiconductor marketing. The semiconductor market can often look like a race to integrate features and functions faster than the competition; that is not usually the case in the consumer world. The marketer of a consumer product such as Coca-Cola (Coke) or Pepsi does not have to worry about integrating more features into the soft drink; nor do they worry about whether their product's functionality meets international standards and specifications, other than those set by the company. Yet, these are issues facing the marketers of semiconductor products. Coke is not being re-designed every day. New features do not have to be integrated into a bottle of soda every six months.

That is not to say that marketing Coke is easy. In marketing a soft drink to consumers, a marketer must understand consumer desires and how to exploit those desires. Marketing a soft drink is more than just advertising; in fact, marketing strategies have to be extremely sophisticated. A marketer must highlight a product's benefits as well as develop a long-term positive image or brand equity for the product. The distribution channel alone for soft drink products can make or break the short-term success and the long-term image established around the product. The brand equity is as valuable as the product itself.

However, semiconductor products use technologies that continuously allow for higher product integration and functionality. In addition to the race for higher integration, there are many levels in the "value chain" that semiconductor companies can target. Commodity products such as memory and discrete components are considered building-block technologies that are low in the value chain but can attract huge volume sales. Highly integrated systems on chip products are considered high on the value chain but typically attract smaller volume sales. Deciding where to play on this value chain is a typical consideration for most semiconductor companies, while it is less a factor for consumer companies.

APPLYING PORTER'S MODEL

Someone who is accustomed to marketing consumer products would find semiconductor marketing frustrating and brutal. Let's use Michael Porter's Five Forces model to analyze the attractiveness of the semiconductor market for semiconductor companies.

* Existing companies: The ever-lasting demand for faster, more-integrated electronic systems and solutions offer ample opportunity to develop innovative semiconductor products. This has created an environment where semiconductor companies continue to develop new products and fiercely compete to maintain and grow their market share.

* Buyers: Large portions of semiconductor products are sold to a few large Original Equipment Manufacturers (OEMs) that dominate their product markets. For example, Dell computer has a lot of power over semiconductor companies that plan to sell product to Dell. (The exception is Intel, which is more powerful than Dell in setting the future of PC product direction.) To sell products to Dell, semiconductor companies have to establish a credible business relationship with the company. These relationships become paramount, especially for small- and middle-sized chip companies. That means that large OEMs wield an influence over the success of most semiconductor companies.

* Suppliers: Most semiconductor companies depend on a few large wafer suppliers to manufacture their products. These suppliers of wafers, and other raw material, have a lot of power over medium and smaller semiconductor companies. Unless a company can develop high-volume successful products, it would have very little power to ask wafer foundries for better pricing.

* New entrants: Various semiconductor product markets continue to attract new companies that enter the market with similar products offering more features and lower prices. This makes it very difficult for any company to have any pricing power. Also, protecting market share proves to be difficult for most companies.

* New product alternatives: In addition, the semiconductor field is the ideal example of Joseph Schumpeter's notion of "creative destruction." There are always new start-ups with new ideas for products that constantly attack incumbents from all sides.

It is this complex world that we call semiconductor marketing — one that has unlimited challenges. It's our goal to simplify this complexity into its fundamental pieces so that all members of a company can adopt a marketing mentality.

A good way to hone marketing insight is through observation. For example, one should take time to observe people's purchasing decisions, marketing campaigns, as well as one's own buying choices and habits. This type of observation provides valuable lessons in how markets work. Many of these lessons are useful for marketing semiconductor products as well.

SUMMARY

Obviously, many elements can affect the success or failure of a product. Everyone should recognize that there is more to marketing than sales and promotion. Marketing is about recognizing or creating customer needs; designing and fashioning products and services to satisfy those needs in a way that complies with the strengths of a company. It is important to understand the power of promotion, pricing, and place of distribution in the overall success of the product in the marketplace. These are the basic tenets that drive the success of products and businesses.

semiconductor industry analysis >

In Chapter 1 we learned that basic marketing starts with the notion of identifying markets, segmenting a target market, and establishing defensible positions in that target market. This chapter provides an overview of the semiconductor industry and highlights how famous companies have succeeded in finding their own defensible positions. Examining how the semiconductor industry has evolved and where it is heading is a good way to illustrate marketing fundamentals at work.

EVOLUTION OF SEMICONDUCTOR COMPANIES

In the early days of the semiconductor industry, finding a defensible position was fairly straightforward. Semiconductor companies in the '60s, '70s, and early '80s simply built wafer-fabrication facilities (fabs) in the United States or Japan and manufactured chips for various markets. The simple strategy of semiconductor companies was to become a large independent device manufacturer (IDM). The IDMs designed, fabricated, marketed, and sold their own integrated circuits. In the early days, IDMs also developed their own Electronic Design Automation (EDA) tools for developing their proprietary ICs. In addition, many IDMs developed proprietary equipment to fabricate, assemble, and test their products. The large IDMs included Fairchild Semiconductor, National Semiconductors, Texas Instruments (TI), Intel, IBM, NEC, and Toshiba to name a few.

These IDMs achieved competitive advantages by vertically integrating all aspects of the design and manufacturing of semiconductor products. In a way, they had no choice — as pioneers, they basically had to build manufacturing capability from the ground up. The vertical structure of these large IDMs also made it hard for many "wannabes" to enter the semiconductor industry. Only a handful of companies had the fabrication, technical, and process know-how to make and sell ICs. This situation led to natural defensible positions for most of the large IDMs.

The early 1980s brought the birth of the microprocessor and the success of the IBM PC. PCs and the development of advanced UNIX servers and workstation, and larger and faster mainframes, led to a need for faster and larger ICs. To meet the need for higher speed and functionality, semiconductor companies started developing Very Large Scale Integration (VLSI) circuits. To address this market, several U.S. and Japanese companies started acting as foundries for companies that wanted to develop pure digital ICs. This was the birth of the Application Specific Integrated Circuits (ASIC) market. The ASIC companies developed their own EDA tools, had foundries, but also started buying equipment from third-party suppliers for the fabrication and testing of their products. The VLSI offerings were limited to mainly digital circuits.

It was during this period that the Taiwanese started making large investments in fabrication technology. Realizing that trying to compete with the large U.S. and Japanese IDMs was going to be difficult, the Taiwanese made a bet that they could make money by concentrating their efforts in pure foundry business. This move gave birth to the concept of the fabless semiconductor industry.

The fabless semiconductor business model gave rise to a whole new set of entrepreneurial semiconductor companies that could now concentrate on creating innovative designs and not be burdened with the large investments required for semiconductor manufacturing. The fabless companies in the late 80's and early 90's could develop several chips, mainly CMOS (Complementary Metal-Oxide Semiconductor) ICs, with just a few million dollars, giving them a competitive edge over IDMs. These companies needed just a few hot designers and creative marketing people to be successful. An early example of an innovative and pioneering fabless semiconductor company was Brooktree Corporation, which used pure CMOS processing to develop the first generation of mixed-signal ICs for the PC industry.

With the evolution of the fabless industry, the formerly defensible positions of the older IDMs in the United States, Japan and Europe were under attack by hundreds of smaller, more agile fabless semiconductor companies. The IDMs were also burdened by the risk and cost of developing new manufacturing processes along with building larger foundries. Fabless companies left this burden up to the foundry companies. In a way, the fabless companies each contributed to the funding of the foundries in Asia to keep up with the latest process technology of large IDMs. The model worked because the foundries could concentrate and focus all their energy on increased manufacturing efficiency; they were effectively able to distribute their investment risk and cost among the many fabless companies using their services.

THE EMERGENCE OF TAIWANESE FOUNDRIES

By the mid 90's, foundries in Taiwan such as Taiwan Semiconductor Manufacturing Corporation (TSMC) and United Microelectronics Corporation

(UMC) had established their business models and were determined to spend as much as needed to keep up with larger IDMs such as Intel, TI, IBM, and major Japanese IDMs. The goal of the Taiwanese foundries was to make sure that large IDMs could not put the fabless semiconductor companies out of business. They did this by ensuring that their process technologies were the most advanced in the world — thus giving their customer base a competitive edge. In addition, the foundries had to build a customer portfolio of products that required large capacity utilization. By building a portfolio of fabless customers who focused on high-volume product lines, they were able to maximize their fab utilization. A good example of the foundries' success is the creation of the Programmable Logic Device (PLD) market segment, which is one of the largest segments of the semiconductor industry.

Today, a battle still rages between IDMs and fabless companies to some degree. However, the lines of battle are becoming more skewed as IDMs start to mix their manufacturing models with fabless outsourcing models. In many cases, IDMs have modified their business plans to concentrate on product markets the foundry companies have yet to penetrate. For example, many of the old IDMs with older process technologies in the United States modified their business plans to focus on creating high-value, low-priced analog products that could serve various end-market segments. These companies used their experience in developing specialty low-cost fabrication and manufacturing capabilities to build a large portfolio of products, each generating a few million dollars in revenue a year.

One IDM that has maintained its leadership is Intel. Intel's manufacturing technology is still one of the best in the world. Intel is also one of the most profitable of all semiconductor companies. It is truly the 600-pound gorilla and in a class of its own. However, in order to maintain its position, it must stay way ahead of the pack in terms of technology and manufacturing capabilities.

Evaluating companies on the basis of their position in the semiconductor industry value chain can also help in your personal investment decisions. While this statement may seem out of context, it really isn't. Investing in semiconductor companies is a great way to get to know the industry. In fact, many of us who work in the industry are already invested in our own companies through employee stock-purchase plans, retirement plans, and stock options. Most of us hope or may actually have faith that our companies have sustainable competitive advantages, which leads to market dominance.

FROM CHIPS TO SYSTEM SOLUTIONS

In the quest for market domination, semiconductor companies must assess where they stand in the "value chain" with respect to their customers. Companies that concentrate on commodity items by definition are working at the low-end of the value chain and typically sell at lower Average Sales Prices (ASP), but make up for it by selling in high volume. Others develop highly integrated products that are expensive and by their nature are used in a specific application. These chips are considered high in the value chain and include ASIC as well as application-specific standard products (ASSP). Due to the high level of integration, ASICs and ASSPs have evolved into Systems On a Chip (SoC).

illustration 5: major semiconductor market segments

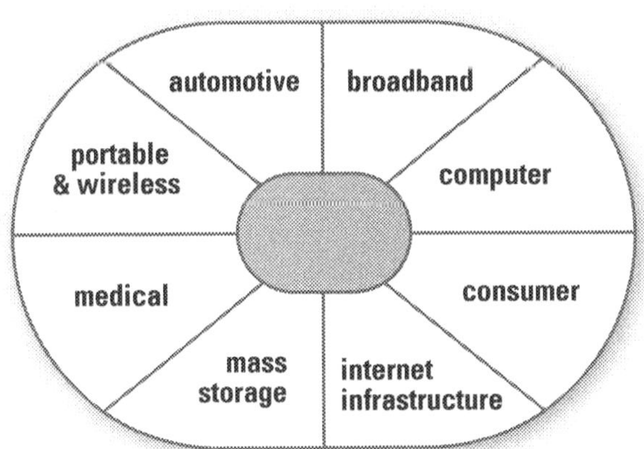

The value chain is important because it can affect a product strategy. Nearly all semiconductor products are used as components to make larger systems. Naturally, if you market chips that are high in the value chain, you must have more end-product systems expertise than a company that sells low on the value chain. For those high on the value chain, product strategy will be linked to the needs of your customers' end markets. In contrast, the macroeconomics of the electronics industry will determine the market for component-level semiconductor products.

Examples of end market segments include personal computers, consumer electronics, automotive, home appliances, military, communications, as well as a growing list of other end markets that use IC chips. In any of these markets, there are a number of large and small Original Equipment Manufacturers (OEMs) that utilize semiconductor solutions and develop end products for various market segments.

For example, Cisco develops and markets networking equipment. Cisco works directly with information technology (IT) departments in a range of industries to determine the attributes of an end solution. It then turns to chip vendors to find the best products for that end solution. These could be both large and small ICs that will be used in end products.[2] Cisco takes on the responsibility of integrating various semiconductor products, developing an operating system and software stack, and finally, manufacturing the product.

A semiconductor company interested in selling high-end ICs to Cisco must spend time analyzing the direction of Cisco's product lines and requirements of its end customers. If the goal is to develop highly integrated SoCs that could influence Cisco's business, then the company must know Cisco's applications and customers as well or better than Cisco. If, on the other hand, the goal is to sell commodity components that have multiple applications, then a less-detailed knowledge of Cisco's products is adequate.

---

[2]In some cases, CISCO also develops its own semiconductors to protect the intellectual property for that product. CISCO does this to help ensure their products prices remain high since by doing this the product cannot be developed and manufactured by others.

Typically, the higher a semiconductor product targets in the value chain, the more support is necessary for the IC from a hardware and software standpoint. At the component level, it is not always necessary to develop software or reference designs to sell a chip. Very low end product such as simple Analog-to-Digital Converters (ADCs), power amplifiers, and other pure analog chips can be sold as stand-alone off-the-shelf products. As one moves up the value chain typically the need for software and reference designs come into play. At the high end, SoCs require a lot of software to be useable. Complete reference designs that integrate the SoC, operating system, and applications software are also required in order to guide customers on how to implement the SoC on their end products.

In the following discussion of the industry at large, consider where in the value chain each company is located and how they have used competitive advantages in their specific markets. The battle to find sources of sustainable competitive advantage at different levels in the value chain is a key feature of the semiconductor industry. The quest for finding competitive advantages drives the strategies of developing, manufacturing, and selling semiconductor products.

illustration 6: semiconductor system solutions

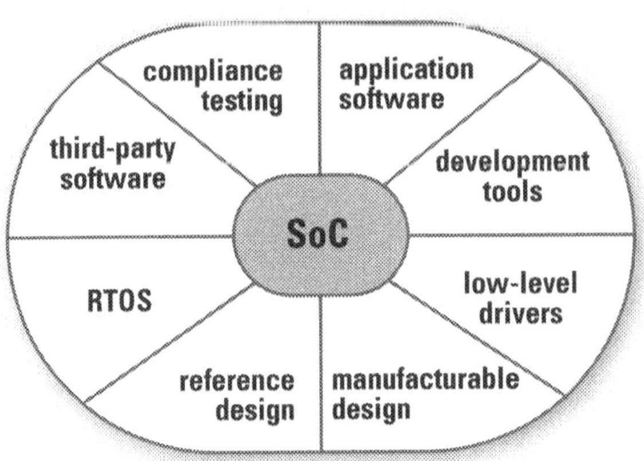

By the way, what is true in the semiconductor industry is also true in the electronics industry. Since the inception of the electronics industry, there has been a continuous battle of positioning between OEMs, semiconductor manufacturers, and software vendors — each jostling for position, each wanting to establish monopolies in their target markets to achieve sustainable revenue and market share growth.

MAJOR SEMICONDUCTOR MARKET SEGMENTS

Thanks to the success of the fabless model and the ever-decreasing cost of semiconductor price performance, markets that use ICs are growing and diversifying. With the popularity of PCs around the world, the emergence of wireless products, the need for more powerful communication products, the increasing utilization of electronic components in cars, the digital revolution of home electronic appliances, and the Internet, there are an amazing number of market segments developing for semiconductor products. These segments include microprocessors, memory, PLDs, DSPs, analog components, and application-specific standard products, to name a few.

illustration 7: semiconductor product segments

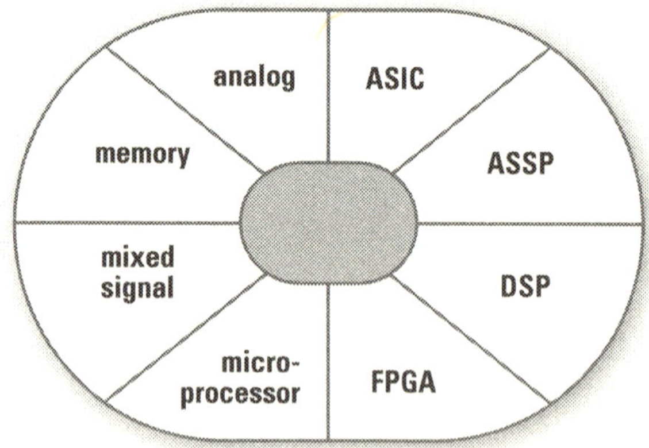

MICROPROCESSORS

Intel Corporation dominates the microprocessor market and is one of the major benefactors of the PC revolution. The personal computer as a system was defined and specified by IBM in early 80's. IBM made a strategic decision to own neither the right to the microprocessor nor the operation system that was brought to them by Microsoft. Instead, IBM concentrated on building the actual system and selling the end product utilizing its brand. The PC market, as we know it today, is dominated by the Intel and Microsoft combination. Through their close work in defining and specifying what hardware and software could effectively be part of the PC, they have created a strong defensible position for their respective companies.

Intel keeps putting more processing power into its chips, while reducing its cost to manufacture those chips. Microsoft develops more advanced operating systems and software applications to utilize all the power available to it. Behind the Intel and Microsoft dominance exists an entire industry of hardware and software vendors that help the two companies get their products to end customers. The main manufacturing and marketing force behind the Intel and Microsoft strategy is Dell Corporation. This means the influence of any single semiconductor manufacturer, except Intel; on the direction of the PC evolution is extremely limited. To make matters more difficult for outsiders, Intel is actively integrating more functionality onto each generation of microprocessor and its PC chip-set.

There are other semiconductor companies besides Intel in the microprocessor market. AMD competes directly with Intel in the microprocessor business. Motorola and IBM provide processor chips for Apple PCs. Interestingly, Apple has recently started using a more powerful microprocessor from IBM to close the performance gap between its PCs and those based on Intel. Some people wonder what would happen if Apple could port its Mac OS onto an Intel platform and compete directly with Microsoft in the operating system and software application markets, eventually becoming a software company.

In addition to microprocessors for PCs, there is also a growing segment of general-purpose microprocessors for embedded systems, such as those developed by companies like ARM, MIPS, and Intel. ARM licenses the intellectual property behind its solution to companies that produce chips for various market segments. Intel is also active in this product segment, developing products for Personal Digital Assistants (PDAs) based on the Microsoft Windows CE operating systems. Dell is now building PDAs based on the Intel and Microsoft solution to compete with companies such as Palm, the maker of the popular Palm Handheld PDA.

Another factor that makes the processor business appealing to existing players such as Intel is that there are a huge number of software applications written by third-party vendors that only run, or best run, on the Intel platform. This means there is an industry of developers who have a vested interest in keeping Intel in a leadership position for as long as possible. This situation helps Intel defend against competitive attacks.

MEMORY

The memory market continues to grow in volume shipments, as software applications demand more and more memory. However, the memory market is a commodity market; the companies in this space have very little pricing power and differentiation advantage. Intel, Microsoft, and Dell heavily influence the standards for memory devices. The investment required to maintain a leading-edge semiconductor manufacturing facility for producing memory products has made many companies turn to their respective governments for financial assistance. The only large American company left in the semiconductor memory space is Micron Technology. To establish a defensible position in this market, a company needs to produce the largest volume, lowest cost, highest density, and highest quality products. Obviously this is very difficult to do. As China enters the semiconductor arena, there is the possibility that companies could transfer their older memory product fabrication facilities to China to benefit from the lower manufacturing costs in that country.

PROGAMMABLE LOGIC DEVICES (PLD)

As noted before, Programmable Logic Devices (PLDs) have several attributes that provide companies competing in this space with defensible long-term positions in the market. The two main players in the PLD arena are Xilinx and Altera. Several factors have enabled the PLD companies to create defensible positions to their businesses. The first item that the PLD giants needed was access to tier-1 foundries that would give them the highest density fabrication process and the lowest cost wafers. As noted above, the PLD business was crucial to the success of the Taiwanese foundries because it provided them with an advanced semiconductor product line. The PLD industry also benefited the foundries by helping the Taiwanese foundries take market share away from ASIC companies, primarily located outside of Taiwan.

The other factor that makes the PLD an attractive sector is that these companies do not have to continuously identify applications for specific product markets. All they need to do is to put more programmable gates on their wafers at a lower cost. PLD companies have continuously benefited directly from Moore's Law, which predicts that the density of transistors on a given die doubles roughly every 18 months at a constant price.

To drive the demand for their products, the PLD companies continuously invest in developing EDA tools that best utilized their solutions. These EDA tools create differentiation and a barrier for end-customers to switch from one company's product to another. It's interesting to note that many fabless companies that spend large amounts of R&D dollars to develop ASSPs, also contribute to the success of the PLD companies by helping to load foundries and drive process technology innovations.

Unlike application-specific IC companies, the PLD companies mostly leave the task of creating product applications to the system companies that use their products. For example, communication and networking companies such as Cisco utilize a great deal of PLDs. Cisco needs PLDs because they give the company the ability to innovate quickly and bring systems to market faster. Even though custom ASICs are cheaper to produce in volume, the development cycles are longer and riskier.

By the late 90's, the PLD companies also had developed various analog cores that system designers could utilize to develop mixed-signal ICs. This trend allowed the PLDs to take some share from the ASIC market. Also, the development of lower-cost PLDs is opening up new market opportunities in the consumer electronics and automotive electronics sectors.

DIGITAL SIGNAL PROCESSING (DSP)

Another attractive market for semiconductor products is the Digital Signal Processing (DSP) market. Companies that currently dominate this sector are Texas Instruments (TI) and Analog Devices. The market for DSP products has grown significantly in recent years due to the growth of mobile phone handset products. Just like PLD companies, the DSP providers do not have to develop application-specific products. Instead, they concentrate on building a development environment that could be used by other companies to develop end products. There are also third-party software vendors that create software products for various DSPs. This means the DSP companies have created an industry around their products that gives them a long-term sustainable competitive advantage in the marketplace.

The DSP companies often focus on specific market segments to better understand the needs of a customer group. For example, TI has established a close partnership with Nokia Corporation, which is the dominant provider of mobile phones in the world. In addition, TI has a large number of DSP solutions for voice and video processing that could be used for various product market segments.

ANALOG COMPONENTS

The analog IC market segment is an attractive market because, in most cases, the products do not need any software to be utilized in end-systems. Hence, chip manufacturers serving the analog IC market do not have to provide their customers with a complete hardware and software applications in order to sell their products. The two companies that dominate this space are Linear Technology and Maxim Integrated Products. These companies have thousands of off-the-shelf products with each individual product representing a small amount of their respective revenues annually. This means that not too many companies can afford to enter this market, because the total available market for each analog product is very small. The key to the success of these companies is that they must have a comprehensive product portfolio, low-cost manufacturing, specialty processes, and a wide network of distributors around the world. These companies also have a large staff of key designers who continuously utilize existing products to quickly introduce new products into the marketplace.

APPLICATION SPECIFIC STANDARD PRODUCT ICS

The Application Specific Standard Product (ASSP) IC companies concentrate on developing standard products for different market segments. These companies face the greatest number of challenges and complex market attributes of any semiconductor product market segment. As previously noted, these products include pure digital ASIC, mixed-signal VLSI, SoCs, and are highly complex devices that need operating systems and application software to make them useful. Unlike the microprocessor market, there are not too many third-party companies that develop software applications for these types of products. This leaves the task of developing a complete software and hardware solution to the semiconductor IC vendor.

There are a large number of standard product companies that compete in this market segment. These companies are not valued as much by the investor community when compared to companies in the PLD, DSP, analog components, or microprocessor markets. This is mainly due to the fact that the markets these companies serve are highly dynamic and often attract many competitors because of the lower barrier to entry. The businesses in this area are both old and new companies such as LSI Logic, Conexant, Vitesse, Applied Micro Circuits Corporation (AMCC), Broadcom, and Marvell Technology Group.

The difficulty with this segment is that the chip companies serving this market have to continuously determine the direction of their target applications to be able to create the right products, at the right price, and at the right time. Although these companies have very smart marketers, it is still difficult to ensure every product is a market winner. Being in this segment of the semiconductor market is similar to being in the movie business — no matter how much movie studios research and plan to deliver a blockbuster movie, success is not guaranteed.

As noted before, ASSP IC product development cycles are long and there exist many external and internal variables that can affect the fate of end products. The growth of the companies in this market segment is often the result of a first-hit product. This means many of these companies end up having difficulty repeating the success of their first product line.

But when an ASSP IC company does achieve a market breakthrough product or can establish a market segment through cutting edge technology, there can be a huge payoff. In addition, if an ASSP company can establish an incumbency in a growing market sector, the payoff can be lucrative and long running.

A primary challenge for ASSP companies is the difficulty of cornering and defending a market segment. Unlike Intel, which has built industry momentum around its product lines, creating a similar model for ASSPs is much more difficult. As will be discussed later, the short lifecycle of ASSPs makes this process even more challenging.

ROLE OF SEMICONDUCTOR FOUNDRIES

It's important to realize that the major foundries wield enormous influence today over their customers. Fabless companies often have a vested interest in promoting certain types of markets and protecting key customers. For example, if a company decides to enter the PLD market, then it might not get the technical and business support needed from either TSMC or UMC, the two largest foundries in the world. Why? Because TSMC and UMC have already aligned themselves with the two largest PLD companies, Altera and Xilinx, and would not necessarily benefit from bringing competition to these companies. On the other hand, if a company has designed a hot new IC for a hot new market not addressed by existing IC companies, the foundries will show more interest in working with that company and may provide it with extensive technical and pricing support.

We can thank the foundries for creating the fabless business model. But being fabless is not the right option in some cases. For example, having a fabrication facility is still necessary if you are developing the latest microprocessor for the PC market, or planning to enter that market. Often Intel adjusts its wafer fabrication processes to develop faster and faster microprocessors, while lowering the product costs. Also, if a company plans to have a complete portfolio of analog component products, then it may need its own fabrication facility. Developing highly complex Radio Frequency (RF) and mixed-signal components requires specialty processes that are not the mainstream business of CMOS foundry giants such as TSMC and UMC. Recently, JAZZ Semiconductor, a spin-off of Conexant Systems, has started providing a specialty foundry service that targets the analog, RF, and mixed-signal IC market.

The fabless model is best suited for companies that have many pure digital CMOS ICs or mixed-signal CMOS ICs. Hence, semiconductor start-ups that embark on developing custom-application specific ICs could benefit the most from working with one of the large foundries in Asia, or IBM in the United States. However, the fabless model may not provide you with the best cost-model if your product does not utilize a significant wafer capacity of a foundry.

FUTURE INDUSTRY TRENDS

The future of the semiconductor industry is as bright as ever. From the macro level perspective, people continue to need products that help them maintain their health, make them more secure, help them to be more productive, allow them to connect and communicate with others, entertain them, and of course, help them make more money. Semiconductor marketers must understand and follow these macro-level needs so that they can design the right semiconductor products.

PC INDUSTRY

During the past two decades, PCs have changed the way we live. The PC is truly a universal machine whose standards are not determined by government agencies. Using a PC, we all have started the digitization process of the content surrounding us. This includes digitization of text, sound, still pictures, and motion pictures. We finally have a mobile office that allows us to work any time and from any place. PCs will continue to be the universal digital content repository and processing machine for many decades to come. And PCs will require higher numbers of input and output devices to allow them to easily connect to the physical world.

Other appliances planning to compete directly with the PC will not be able to attract the critical mass of users and financial capital they need to survive in the long run. This means that companies such as Intel, Microsoft, and Dell will continue to dominate the PC industry and with their power will influence the direction of all semiconductor products that are and will be designed for the PC space.

COMMUNICATIONS INDUSTRY

Mobile communication devices such as cell phones will continue to evolve throughout the next few decades. The industry will add better processing, transceiver, display, and battery technologies to a growing array of mobile devices. Large mobile phone companies such as Nokia and their service-provider partners will continue to shape the market and guide the direction of the mobile device evolution. The semiconductor players in this market must pay attention mainly to companies such as Qualcomm that have the Intellectual Property (IP) behind the transmission standards. In addition, DSP companies such as TI will continue to be a leader in providing the industry with a unified DSP product line, which will play an ever-increasing role in mobile phone applications.

AUTOMOTIVE INDUSTRY

During the next two decades, the automotive industry could become one of the largest users of semiconductor products. The need for Global Positioning Systems (GPS), DVDs, game consoles, and high-end automotive audio systems will continue to grow rapidly. In addition, electronics are added to cars to help improve safety and engine performance. We will see a growing need for ICs to help deliver the security, communications, performance, and entertainment applications for tomorrow's cars.

NETWORKING AND TELECOMMUNICATIONS INDUSTRY

Both consumers and corporations continue to depend on digital connectivity to improve productivity and communication. Hence, the need for networking and telecommunications equipment will continue to grow. We will eventually be able to see and hear anyone at any time at some point in the next two decades. This means that companies such as Cisco will continue to have a healthy business for many years to come. To re-ignite the networking equipment market, we need to see various governments making broadband Internet access as accessible in homes as electricity, phone, and water is now. In addition, as PCs continue to make personal video conferencing easier, we could finally see the emergence of video conferencing at the corporate level, therefore driving the need for more networking equipment.

In the short term, semiconductor companies operating in this space are experiencing a sharp downturn for their products. This is due to the over-building of the networks during the dot-com era, which was financed primarily by financial markets and not actual service revenue. The role of the system provider, such as Cisco, in this space is of great importance to semiconductor companies working in this market.

CONSUMER ELECTRONICS

Digitization of different content such as text, audio, and video will contribute significantly to the consumption of semiconductor devices in consumer electronic appliances such as digital cameras, DVDs, high-definition TVs, digital audio players, etc. In addition, with the success of high-end gaming units (such as the Sony PlayStation and Microsoft X-Box), there is an opportunity to have more silicon content in home electronic appliances.

Large consumer electronic companies, with global brand names and comprehensive product portfolios will continue to control most of the profits in the consumer electronics business. Companies such as Pioneer, Sharp, Panasonic, Sony, Samsung, and Philips continue to innovate in this area by developing new standards that will accelerate the process of digitization of all content that is received, stored, and used in the home. We see the PC moving from the home office to the living room and becoming the digital media receiver and storage center of choice. Of course, these home digital media receivers will not look anything like the work PCs of today.

These digital media servers will also serve to store content from a host of new digital products such as digital cameras, digital camcorders, color printers, DVD players/recorders, and high-end audio systems. Together with the possibilities of the internet, consumer electronic companies will continue to innovate and bring new bundled products to market.

Apple Computer is trying to take advantage of these trends by producing various consumer appliances that will extend the usefulness of its personal computers. Apple iPod is an example of successful consumer electronic product that is allowing the company to have revenue stream after the initial sell of the product, by bundling the iPod with music download service as well as managing digital content on the computer.

In addition to consumer electronic appliances that do not need a service provider to make them useful (such as DVDs), cable and satellite set-top boxes continue to keep pace with the needs of consumers. Cable and satellite service providers will continue to control the set-top evolution as they view the hardware to be an extension and part of their network. For example, in the United States, EchoStar (the company behind the Dish Network) and News Corporation (the company behind the DIRECTV network) will control the form, fit, and functionality of their respective satellite receivers.

We also believe that consumers increasingly use their PCs and game machines for digital content creation, storage, and delivery. As was noted, Sony and Microsoft are evolving their game solutions to compete directly with the functionality that set-top box and entertainment service providers want to add to their systems. In the future, we will see delivery of IP-based video content into homes using broadband pipes and game appliances that will directly compete with the service offerings of both the satellite and cable operators.

Semiconductor companies that develop long-term strategic partnerships with consumer electronic appliance makers will continue to reap benefits of this product segment. For example, LSI Logic provides the processor for the Sony PlayStation 2 game console. The PlayStation 2 easily could have a four- to five-year lifecycle, giving LSI Logic a predictable revenue source. Also, Intel and NVIDIA provide the microprocessor and graphics controller used in the X-Box, giving those companies an opportunity to be part of a new breed in the home game and entertainment appliance market.

CONSUMER BROADBAND AND HOME NETWORKING

As the need for residential broadband access and home networking grows, we will see a host of new electronic appliances that allow consumers to easily access the Internet and manage their home network. In addition, the emergences of smart home appliances and access to IP-based media content will drive the need for higher-speed access to the Internet. Many governments, such as Korea, are mandating that broadband service providers offer broadband access to all homes by the end of this decade. We would like to see more direct government involvement in driving the penetration of broadband access in the United States and Europe. This would have a significant impact on growing the need for semiconductor devices.

As consumers increasingly access the Internet using a broadband connection, the need for home networking and access sharing grows as well. This gives both the consumer electronic manufacturers and PC manufacturers the opportunity to develop consumer appliances that benefit consumers and allow the free flow of digital content between home devices. The recent explosion of home networking using the IEEE 802.11x standard is proof that the home-networking market has a great future.

Today, the primary technologies that consumers use for broadband access to the Internet are DSL modems and cable modems. Telecommunications companies and cable MSOs (Multiple Service Operator) will influence the evolution of these devices. DSL will continue to be the primary broadband access throughout the world, due to the ubiquity of phone systems. We also see the need for developing the fixed and wireless broadband-access markets. This will bring much-needed competition to the wired broadband-access market.

Home-networking standards will continue to be influenced by IEEE standards committees. Consumers will likely decide which home-networking standards will win out over the long term. It is still unclear whether home networking will be dominated by wired or wireless technologies — and it may very well be that both will be used for different applications. The key driver for adoption of new home networking technologies is consumers' need for having a robust home media network.

As noted before, the market for home-networking and broadband-access devices in the United States could grow rapidly if government officials see how far behind the United States is in the deployment of broadband access to homes. We believe that mandating broadband access for every American home within the next five years could be a great catalyst for re-igniting the U.S. economy.

SUMMARY

As they have during the past four decades, semiconductors will continue to provide technologies that will find their way into millions of applications. The universality of the digital revolution will continue to drive the need for more semiconductor products. The industry is adjusting to the new realities of the networking demand, as public and private market funding for big capital expenditures is also readjusted. The phenomenon of companies vying for defensible positions will continue to drive innovation and establish new business models.

The various industry segments provide an array of opportunities and threats for both existing and new entrants into the semiconductor industry. As is discussed in the next chapter, it's important for companies to do a thorough analysis of their capabilities before targeting any product segments. It is necessary to have an accurate assessment of where in the value chain you can or should be to achieve a sustainable competitive advantage. The lifecycle of semiconductor products continues to get shorter, while the complexity of developing new products increases. This means that anything but perfect product development and execution can severely damage the long-term prosperity of most semiconductor companies.

developing a marketing strategy ›

In chapter 1 we learned that there are three basic steps to marketing: identifying markets, targeting markets and positioning products and/or services for those markets. Now let's see how these principles apply to the semiconductor market and let's see how the initial stages of a marketing strategy are developed.

Imagine that someone just gave you ten million dollars and ordered you to start your own semiconductor company. How would you go about deciding which market to enter? As we know from chapter 1, your first step would be to identify a promising market. There are many markets open to semiconductor companies including microprocessor, memory, analog, PLD, DSP, ASSP, and ASIC, among others. In addition, there could be entirely new markets to pioneer.

Narrowing down the options may be easier if you already have an understanding or knowledge specific to a market. A start-up or established company bases its decision on what market to serve by applying a set of core competencies or strengths to what it believes is a market opportunity. For example, it is likely that a group of RF engineers would use their core competencies to create an RF product company. An established IDM may decide to start a foundry services division to help load their fab, if they believe there is demand. So, if someone gave you ten million dollars to start a venture, the first step would be determining what your core competencies are.

If you had no core competencies and you are holding a check for ten million dollars, your set of market alternatives is wide open. This may be a blessing because it allows you to focus on the hottest market opportunities or lets you make a bet on an entirely new opportunity.

APPLYING SWOT ANALYSIS

In Chapter 1, we briefly discussed the SWOT (Strengths, Weaknesses, Opportunities, and Threats) analysis as a way to help target market segments. We also introduced Michael Porter's approach to determining market attractiveness. These tools are useful to help determine which market segments to serve.

Let's use SWOT analysis to various semiconductor market scenarios to see how the dynamics of SWOT and market segmentation can come into play. For a simple example of opportunity versus threats lets consider the semiconductor products that make up Microsoft's X-Box. Looking at the Bill Of Materials (BOM) for the first generation of X-Box, we see that the two major silicon components of the box are the microprocessor (Intel) and the graphics controller (NVIDIA). There are also passive and active components in the X-Box, in addition to a few other silicon components. So, for example, there may be a great opportunity to provide a high-end, cost-effective processor product for the next generation of Microsoft X-Box game console. The processor solution for the X-Box enjoys a healthy Average Selling Price (ASP) as well as significant volume. It is no doubt this is a great opportunity and much better than a lot of the other component-level silicon in the box.

illustration 8: SWOT analysis

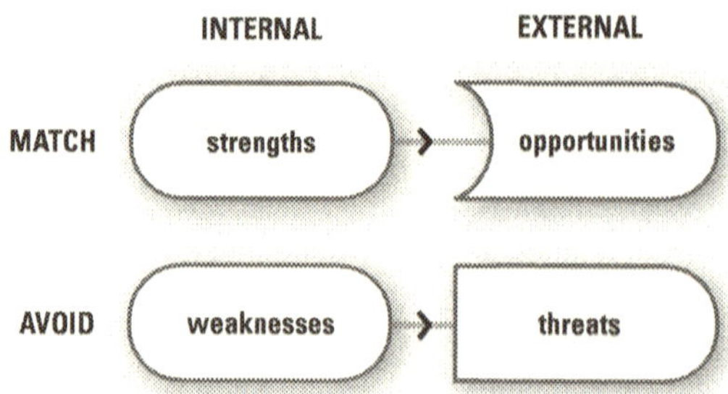

A company might decide to go after the microprocessor segment of the X-Box if opportunity were the only deciding factor. However, the huge opportunity for designing a microprocessor for X-Box might be offset by the threat of not being able to convince Microsoft to switch from Intel to an alternative solution.

THE IMPORTANCE OF SWITCHING COST

The threat is not limited to just Microsoft's decision. Intel, being the incumbent supplier, can simply lower the price of its current, as well as the next generation, processor in response to your attempt to enter the market. There are huge marketing challenges associated with unseating an incumbent due to costs associated in a switchover. In the case of the X-Box, there is a community of third-party game developers that are, or will be, supporting the Intel-based X-Box game consoles. Any move by Microsoft away from Intel would incur not only hardware design costs, but also massive software switching costs. Even if the competitive solution can offer chips at less than half the price, the switching costs could easily outweigh any potential savings per chip. The case for having Microsoft switch from Intel to a new microprocessor for the X-Box is very difficult; the threats clearly outweigh the opportunities (at least currently).

The problem of switching costs is a major challenge (threat) affecting many SoC market segments. It is a challenge to the non-incumbents and a great form of competitive advantage for incumbents. A challenger must invest a huge amount in developing a complete software and hardware solution to overcome the problem of switchover costs. Even with this level of investment, the non-incumbent also needs to overcome any perception of risk associated with a switchover. A challenge only makes sense if prospective volumes and projected ASPs are enough to justify the cost of entry.

The lower a semiconductor product is on a value chain (for example an RF component), the lower the switchover costs. However, there are threats lurking at the low end that are just as menacing. On the commodity off-the-shelf component level, price is a major determining factor in getting design wins. Often, a reasonable savings on the components' cost justifies a switchover to a competing solution, assuming the performance and quality of the product are as good as or better than the incumbent's product. However, it is unlikely that a company will justify a redesign to take advantage of the lower price of a single component. Most often, customers switch to new components only during system re-designs or the cost-reduction phase of a product's lifecycle. That means if a new component could save Microsoft on the cost of the X-Box, it must be ready when the X-Box cost-reduction window of opportunity opens, if there ever is one.

PREDICTING THE THREATS

Since developing a commodity component is not normally difficult, one problem with the component market segment is the potential number of competitors that could easily enter a specific market segment. These competitors will always pose a threat to any position one might establish. There may be some exceptions to this, such as exotic high-speed RF components used in mobile handsets. However, for the most part, the commodity business is by its nature a crowded, cutthroat segment.

In addition to the number of competitors, one should not ignore the market power of competitors that offer large product portfolios. Unless a new entrant plans to introduce a broad product portfolio, larger vendors that can offer cost savings across a range of products will normally have the advantage. To establish an attack against competitors with large portfolio offerings requires massive investment.

The components business is also under threat by the SoC vendors that are continuously integrating the smaller components onto their new SoCs. This is a threat that is particularly acute in the PC space. Component vendors can enjoy opportunity for only a limited duration. For example, take USB 2.0. There is a market window to provide USB 2.0 as a separate chip. But it is only a matter of time before the USB 2.0 technology is integrated into PC chipsets.

In fact, recent announcements by Intel indicated that the company might eventually integrate various external wireless LAN components into its

PC chipset. There is also the threat both Microsoft and Intel continuously want to move various processing functions away from custom IC products into its OS. A good example of this is in the area of video encoding and decoding SoCs.

IDENTIFYING OPPORTUNITIES

Non-PC market segments, such as consumer electronic products, also offer their own set of opportunities and threats. However, unlike the PC market segment, establishing important partnerships with non-PC system vendors is much easier due to large number of non-PC system solutions developed every year. Establishing strategic relationships with service providers, consumer electronic brands, and other OEMs is the key to success of any company targeting the non-PC segments of semiconductor market. The goal is to determine the required features and functionality of ICs that system vendors need ahead of competitors. This applies to both incumbents and non-incumbents.

Success in the ASSP market requires technology and product innovation that matches windows of opportunity. These windows usually occur during a technological shift or market discontinuity. For example, in the wireless LAN space there were existing players that provided 802.11b complete IC and software solutions. Since everyone was demanding higher bandwidth, new entrants into this market only had opportunities if they could come out with newer 802.11g, 802.11a, or other variants of the technology that markets demanded.

A semiconductor marketer must spend significant time and effort to determine true market opportunities that match his company's strengths. In addition, by segmenting the market further in terms of value chain, the marketer needs to further analyze the threats associated with the segment. Ideally, segmentation should be done consistently across many product lines in order to create synergies between families of products, and to leverage the entire product portfolio of the company These synergies across product lines are key differentiators that can help position a company against companies with a single product offering.

Market segmentation can get very involved. An expert marketer may segment a market many different ways and to various degrees of granularity while going through the SWOT analysis to identify target markets. For example, a market such as set-top box can be further refined to satellite, cable, and terrestrial set-top box markets. These segments can be even further refined to a myriad of additional segments.

The goal of SWOT analysis, and extensive market evaluation, is to target the most promising segment. When the SWOT analysis is done correctly the segment chosen will already have the seeds of sustainable competitive advantage planted. Once a market is targeted, the next step is to undertake product positioning decisions that further bolster competitive advantage. Competitive advantage is what keeps Microsoft and Intel executives awake at night. Technology is always changing; new wrinkles in the technology landscape can quickly turn today's giants into tomorrow's has-beens. This is the reason why Andy Grove of Intel coined the phrase "Only the paranoid survive."

SOURCES OF SUSTAINABLE COMPETITIVE ADVANTAGE

There are three strategic focuses at a marketer's disposal to position a product (or more generally, a company). These are product innovation, manufacturing innovation, and organization innovation. Effective use of these strategies establishes sustainable competitive advantage for a product or company.

Many executives believe that their companies must deliver the best in class on all three fronts to be able to survive and have a real chance of defeating competitors in the marketplace. However, this is rarely the case. Most companies excel at one or two areas but rarely, if ever, at all three. Intel is a leader in product differentiation, but not necessarily in price. Southwest is a price leader, but its planes are the same as everyone else's. One could argue that Dell is a price and, to some extent, a service leader, but not necessarily a product differentiation leader. The point is that excelling in just one or more strategic area is good enough to establish a strong position in a market.

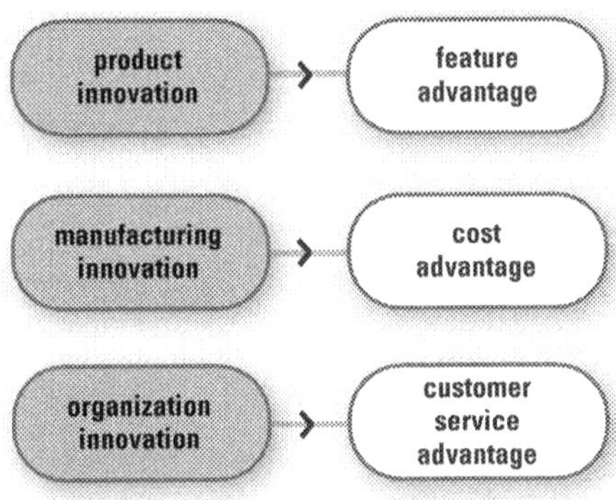

Let's look at each strategy and discuss how each can contribute to establishing a sustainable competitive advantage.

PRODUCT INNOVATION

This is the main strategic element that most semiconductor companies pursue to create a long-term defensible position against their competitors in the marketplace. This strategic thrust greatly depends on how promptly a company can innovate new products that a market demands. True innovation comes from observing market demands and then delivering a compelling solution that competitors have a hard time matching. An example of this is Texas Instruments' innovation around its Digital Light Processing (DLP™) technology. TI saw a growing market need and trend for higher-quality light-processing technologies that could create a new class of portable projectors. In addition, the same technology is now being used in HDTV-ready rear-projection TVs. This product innovation has given TI a product position that would be hard to match by any new entrant into this market segment.

Creating a product differentiation strategy requires a highly educated staff. Companies that are following a differentiation strategy must establish and maintain ties with tier-one engineering schools that could provide them with a constant flow of fresh ideas and talent. In order to attract this level of talent, a company must have a reputation for innovation. Therefore, the reputation of your company not only affects your customers' perception, but the perception of the highly skilled workforce you must attract to create innovative products. Having the top talent and being able to attract the new top minds in technology to a company is a great source of sustainable competitive advantage for that company.

Often, start-ups that establish a business based on a single innovative product idea expect to continue their success with the original group of people that created the first product success. However, depending solely on a few key people for future product innovations is dangerous for obvious reasons. When talented people leave, so does a company's future. Therefore, innovation requires a constant dedication to hiring and maintaining the best talent available.

illustration 10: success factors for an innovative product strategy

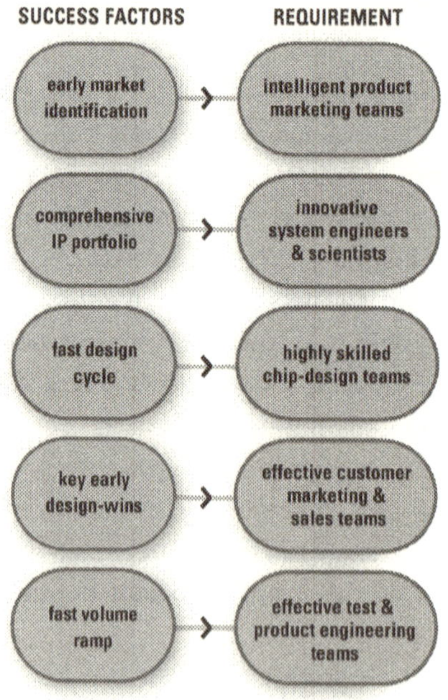

A company that is driven by innovation must keep tabs on its talent pool by counting the number of engineering Ph.D.s working in development groups and the number of fruitful patents filed each year. Also, a company must separate how much current revenue is from products released within the last three years and how much of it is from products released beyond three years. Innovative companies constantly study these product revenues trends. If a trend shows that most revenues come from legacy products and not new products, it points to serious problems. It could be that there is degradation in product innovation, which could lead to losing talented staff to other companies or cause a drop in revenues.

A company whose strategy is to be an innovator should not obtain all of their IP needs, design know-how, and manufacturing needs from outside the company. In fact, the semiconductor business requires that a certain amount of in-house capability exists. A semiconductor product company will not likely survive in the long term if it does not create in-house technologies and defensible intellectual property. This is because a company that loses its ability to innovate will also eventually lose its position as a low-price leader or as a service leader. This innovation does not necessarily have to happen at the product level. Manufacturing and customer support innovations are the key to long-term success of any semiconductor company.

Even Intel's aggressive brand marketing is backed by product and manufacturing innovation that allows its customers, such as Dell, to rely solely on Intel for its microprocessor needs. It may be that the short-term easiest path to fill a product or technology gap is to obtain it from outside, but in the long term, this would lower the competitiveness of internal product development groups. Instead of pushing hard to innovate and create new products and technologies, the company would end up with a group of people acting as consultants that just evaluate any semiconductor IP.

The perception of having innovative product in the mind of customers results in tremendous competitive advantages for a company or product line. A reputation for innovation results in the ability to command higher prices for a product. Customers will always be eager to consider an innovative company's technology for their designs. In addition, customers feel safe and secure to stay with an innovative company's products, and will call that organization first for their future needs.

MANUFACTURING INNOVATION

Another primary tool companies can use to entice customers to switch to their solution or to maintain their existing customers' business is to lower the selling price of their products. However, this only works over the long run if a company or product line has a sustainable manufacturing innovation, leading to cost advantage over the competition. After all, lowering the selling price without having a cost advantage will result in lower operating margin. This issue is especially important from a financial reporting aspect for publicly held companies. If a company has set an expectation to Wall Street to achieve a certain operating margin, it must maintain or increase that margin expectation. Analysts who represent the investment community get concerned when they see operating margins dropping; needless to say, this can have a negative impact on a company's stock price.

Companies offering commodity products are often faced with a large number of competitors, and usually organize their company around a cost-advantage strategy. They depend on having a dominant market share, lower fabrication cost, lower packaging cost, and lower test cost in order to maintain a healthy margin for their products as prices drop. For example, semiconductor companies based in Taiwan have lower operating margin requirements than those based in the United States. This means they can effectively compete in the marketplace for commodity products by having a lower average selling price.

Having a low-cost strategy in the semiconductor product business requires access to low-cost manufacturing. As previously noted, factors that determine semiconductor product costs are mainly wafer fabrication, assembly, and test costs. If a company depends on external foundries, assembly, and test houses for the bulk of its production needs, then the company must secure competitive production pricing from these vendors to execute its cost-advantage strategy. For example, if a company has a product line that is manufactured at TSMC and plans to compete in the marketplace by having the lowest ASP, then it must have very low wafer cost from the fab — assuming the company is determined to deliver the same gross margin as its competitors. TSMC offers volume discounts to companies that have volume products, or if there is some strategic reason to extend favorable pricing. A company that cannot secure low wafer pricing should not embark on a low-cost strategy. Without a low wafer cost, a company must depend on the innovative capability of its design team to provide a small die size, using the latest process geometries, to achieve a lower die cost.

Companies that depend entirely on a low-cost strategy must couple low production costs with a design-for-cost mentality. Products must be designed to have the lowest cost target possible, given a set of features. Semiconductor ASPs only decrease in the long term as new competitors vie for market position. It may be easy to low-ball a price to win a design, but the business can only be sustained if costs follow suit.

Cost reductions for SoC solutions must also be gauged from a total BOM standpoint. Your customer will not switch from an incumbent solution to your solution if it results in a higher system BOM cost. In addition, the SoC solution must represent a smaller switchover cost. Therefore, being a cost leader in the SoC arena requires cost-reduction strategies on multiple fronts.

It is not easy to be a cost leader in the semiconductor market. It requires a company to focus on cost-reduction activities in almost all aspects of design and manufacturing. A low-cost strategy must be part of a company-wide focus. It must include having very little management overhead, no executive luxuries, and a world-class low-cost operation engine.

ORGANIZATION INNOVATION

Good customer service can provide companies with competitive advantages that are hard to beat or duplicate. But as was the case with cost strategies, successful customer service also depends on a company-wide focus on creating a highly agile organization. If everyone from a company's CEO to its most recently hired junior employee sees the customer as the central focal point of all activities, then customers will view that company as one that delivers world-class customer service.

A customer service strategy directly affects the long-term success of many companies in low-tech industries. For example, customers rate airlines depending on the level of customer service they provide and their safety ranking. It's true that we all want lower airfares, but we also want to be treated with respect and courtesy by an airline's employees, and we certainly want to survive the trip.

Selling semiconductor products requires heavy engagements with customers during and after the initial sale of products. During the pre-sale phase, customers require constant attention to ensure their designs are done right. Customers depend on the application engineering staff and sales team to give them individual attention. They also need special attention once their products are released to their target markets. They want vendors who understand what is needed to compete against products using alternative semiconductor components. If customers do not get the attention and service they need, they are more likely to switch to a competitor. Great customer service, coupled with a positive attitude, can't easily be duplicated.

illustration II: customer focused company

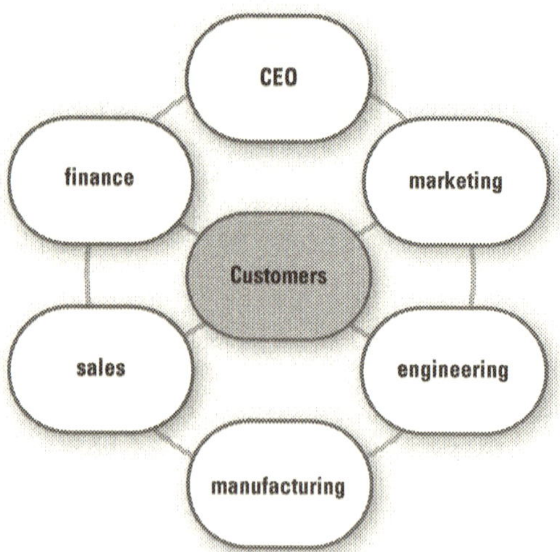

Establishing a company totally dedicated to customer service is also very difficult. Every stage of product development must take into account providing the best service to customers. At the design level, one can start to consider how to make products that are easier to support from an applications standpoint. Of course, planning market-appropriate hardware-reference designs and well-written software documentation is also critical.

Let us reiterate that it is extremely difficult to deliver world-class innovative products, at the lowest costs, and with the best customer service for all your product lines. A company should decide which one or two out of three factors they want to stress. It should then set the right product positioning given the skill set and assets of their organization.

Once a market segment is selected, a marketer must decide how to position a product line to establish a defensible position. The three ways to do this are through product differentiation, price, or a customer-service strategy. Ultimately, how one decides to establish a defensible position depends on a company's inherent strengths and weaknesses.

In addition to the strategic decisions outlined above, a final word is applicable regarding the eventual success of whatever path is chosen. Regardless of which strategy is adopted, successful companies or business units translate their strategies into a sense of mission among the employees. It helps also to have a little competition so that the image of the "external enemy" can foster a competitive spirit among the rank and file. Having or creating an external enemy is a great way to create this focus. "External" is the key word, as creating the "internal enemy" is the kiss of death for most companies. The real enemy should always be an external company or set of companies. The external enemy, coupled with a company's own clear vision, mission, and strategy can turn businesses into aggressive, successful organizations.

Successful CEOs monitor and modulate the attitude of the entire organization and the senior staff to make sure that everyone in the company develops a "can do" attitude. A company works best when a clear vision from the CEO streamlines with individual marketing campaigns of various product lines.

Younger companies seem to do a much better job of helping their people realize what they have to do to beat competitors. The challenge is bringing that winning attitude to an older organization. By its nature, a semiconductor company depends heavily on the skills of the people who develop, market, sell, and service the products. This means that, without the right attitude, companies can quickly lose their lead in the marketplace.

SUMMARY

Deciding on a market strategy is more than just picking the market with the greatest revenue opportunity. The analysis requires some deep thinking about the right market segment, coupled with a solid product strategy based on sources of sustainable competitive advantages. A marketer must align his company's strengths with market opportunities as well as determine the level of threat for each market segment under consideration.

Finally, whatever strategy is selected, the chance of success is increased considerably if the strategy is supported from top down. The CEO needs to foster a sense of mission and urgency around the strategy and motivate everyone to fight for success.

A key element to implementing a successful semiconductor strategy is the development of the roadmap. Semiconductor companies are business-to-business ventures with products that have limited lifecycles. Therefore, semiconductor companies must sell themselves to their customers as viable vendors for the long haul. This is especially true for companies in the SoC business where software investments can be huge. Customers want to partner with companies that show they have a viable long-term roadmap and a plan to successfully execute their business objectives.

Road mapping helps semiconductor companies define their product strategies for various market segments. It also helps in establishing development plans and long-term strategies for follow on and future products. Marketers rely on roadmaps to communicate their vision and plan to customers, management, and engineering.

ALIGNMENT WITH COMPANY VISION

A product roadmap should ideally reflect the vision and strategic direction of the entire company. In many cases this vision and direction are not well communicated to marketing and engineering teams at the working level. And many times the CEO does not even have a vision to communicate, or more often, the vision and direction are unclear or unexciting.

Whether or not a company's leadership is strong, a marketer needs to consider himself the CEO of his product line. Even though he may not be in charge of engineering resources or responsible for overall profit and loss of the product line, that person must behave as though he lives and dies by the success of his products in the marketplace.

Much like a CEO "selling" his business plan to investors, his company, and his customers, a marketer must also sell his roadmap. The marketer must sell his roadmap to customers, the senior staff, as well as the engineers. Developing, promoting, and executing on a solid road-map is fundamental to semiconductor marketing success. Think of it as your product line's business plan expressed in a set of slides or via a written document.

In practice, the roadmap can be one slide showing the products available today and a set of future products that offer feature enhancements, cost reductions, or other benefits critical to a target market. More likely, several slides comprise a roadmap. Whether it's one slide or one hundred, a roadmap presentation has to communicate enough information for someone to buy into a product line. The roadmap must sell a product vision, and more importantly, convince customers to buy chips and align their product planning to the roadmap.

ROADMAP DEVELOPMENT PROCESS

Unlike engineering, marketing is not an exact science. The process of creating a roadmap is not black and white. In fact, product roadmaps are hotly debated within an organization. In addition, unlike an engineer who might discuss a design strategy or architecture with a few colleagues, a marketer is forced to debate his or her strategy with a variety of people, from the CEO to the engineers. This is the way it should be; marketing decisions have direct consequences to the bottom line.

As previously stated, the roadmap has to be sold not only to customers but also internally to company decision-makers. It should have solid backing by as many decision-makers as possible. The owner of a product roadmap must evangelize it so that people become convinced of the plan and disciples of it. If a marketer is not confident in his own plan, he is more likely to change it constantly depending on short-term market events, or other people's input. A constantly changing roadmap has marginal value and will detract from the authority and confidence in the marketer in charge of the product lines. A marketer must develop self-confidence about his market and product offering, and give a true assessment of market realities and the future of the market direction.

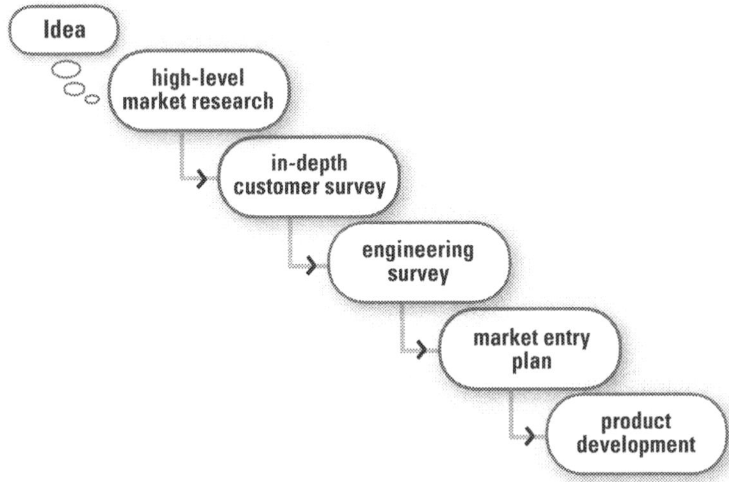

Generally, new roadmaps are generated following four important steps: market analysis, customer surveys, competitive assessment, and conducting internal company promotion and buy-in. Roadmap generation for existing product lines are discussed later in this chapter.

STEP 1: MARKET RESEARCH

The first step to effective marketing is having a deep understanding and appreciation for a target market. By the time a good marketer starts to put a roadmap together, he will have gone through an in depth analysis including market identification, segmentation and targeting. At this point, it is assumed that the product positioning, definitions, etc already are utilizing the company's sources of sustainable competitive advantage.

Roadmaps are plans that show product evolution over a specific timeframe. Depending on the type of semiconductor products that timeframe can be anywhere from six to 60 months. At this stage, product ideas are still untested and a roadmap should be considered as a first pass or a "straw man" product roadmap.

STEP 2: CUSTOMER FEEDBACK

Once the straw man roadmap is completed, it has to be taken out and shown to customers and other decision makers. Customers provide valuable feedback through their questions, comments and criticisms. They are eager to discuss their problems and the solutions they need.

Marketers should invite their key engineers to attend these customer roadmap reviews. Through customer meetings, engineers can hear firsthand (and with presumably more technical insight) what customers are asking for. They can also ask questions and raise issues in real time. It also stimulates engineers to think about how they can implement solutions. Attending meetings with customers can get engineers excited about prospective designs. Finally, when it comes time to sell a roadmap internally, engineering will be in a better position to evaluate marketing proposals.

Customer feedback is crucial for ensuring that products will be relevant in the market once they become available. In addition, customer feedback is necessary as justification for ideas when selling a roadmap to senior management. Internal management and engineering will always put heavy pressure on marketing groups to make certain their product roadmaps are real and "doable" (and not just paper-based product roadmaps).

Interestingly, it is the authors' experience that customers like to advise companies on how to enter a market segment. System vendors often invite chip vendors to their review and learn about their end product roadmap, since it is to their advantage from a supply and cost standpoint to have chip suppliers competing for their business early on. Customers will often perceive new entrants as someone with potential — someone who, if successful, could bring a new set of needed products to that sector. Customers are very likely to offer a lot of advice about what they need from a chip company to make their end product line successful.

In the early phases of determining a roadmap, marketers should make certain they point out to customers that plans are still in flux. Most customers understand that marketers are often in the process of canvassing the market in order to solidify product plans. It is important in these early stages to interact with friendly customers who seem eager to promote new technology. Marketers should put off approaching tougher customers until a roadmap has matured.

Customer interaction is critical, but relying too much on customer feedback can also be problematic. In general, customers are focusing on short- to mid-term needs, so it is always necessary to balance what customers are saying with market analysis and long-term goals. Also, keep in mind that some customers may have divergent views on a product roadmap and product features. A good marketer must decide what path to take based on his own business analysis of the market direction. Finally, if a marketer's technical and marketing insight is strong enough, it is possible to direct and shape customers' plans by presenting what is possible. In effect, a comprehensive marketing effort can drive the direction of a market.

Roadmaps are easy to develop — building a successful product is hard. It is easy to over promise technology and deliverables to customers during roadmap discussions. Too much over-promising can lead to loss of credibility in the long run. Marketers must develop a deep understanding of what it takes to execute a product roadmap by examining all external and internal factors.

The scarce resource in developing a successful product line or a new company is not financial, but human. In a sense, the ability of a technical team is the key differentiator in gaining a true competitive advantage that can't easily be reproduced by competitors. Many experienced people in the industry often remind us that it takes twice as much money and twice as long as planned to bring a product from conceptual stage to the revenue-generating stage. If that is the case, then marketers must set the necessary buffer in their roadmap to give customers a realistic timeframe about when various products featured in roadmap slides will be available.

Therefore, it is a good practice, especially in early customer visits, not to give them exact product-availability dates. After all, schedules often slip. Engineering teams can give in under pressure to unrealistic schedules from their marketing teams. This creates a situation where a marketer's wishful thinking can produce an unrealizable schedule that leads to losing credibility with customers.

Conducting a good customer survey can take some time. Just like finding a good car or a house, it is a good idea to shop around for as much feedback as possible. A product roadmap is often the result of several months of studying various market drivers, competitors, and customer needs, and then designing a set of products and solutions that can be developed and sold with a healthy operating profit.

STEP 3: COMPETITIVE ASSESSMENT

After the roadmap starts to take shape, it is necessary to do a complete review of the competitive landscape. This exercise requires courage. A good marketer must take in the competitive landscape and not only ensure that his new product will beat the competition's products today, but will also be better than what the competition can produce in the future.

Marketers often over optimistically discount the competition. It is best to assume that the competition is working very hard and is considering the same obvious moves as everyone else. It can often happen that a marketer will specify a chip with more features and integration than what the competition has in the market today — only to be surprised to find that the competition introduces a better chip later on.

Any kind of competitive assessment will require taking the longer view. Depending on the chip, development to product roll out can take anywhere from one to two years. A marketer must consider where the competition will be in that time frame and determine what features, integration and capabilities will beat the competition at that time. If the competition is unaware of the coming attack, perhaps they will make the mistake of relaxing. However, most semiconductor companies are battle hardened and paranoid. Therefore, they will be on the look out for new competition.

A brand new product line's roadmap should also specify a market entry strategy. This entry strategy must take into account timing, promotion, positioning as well as sales channel strategy. A determination of the value products bring to customers is necessary to determine sources of competitive advantage. This is especially important for winning those customers who have alternative suppliers.

STEP 4: INTERNAL SURVEYS

Engineering can and should provide a great deal of help in the development of a roadmap. Involving engineering is essential in order to get buy-in for a product line. Ideally, marketers should seek the buy-in of a senior, well-respected engineering manager, or a system designer, who is recruited to help them champion a product line. An engineering champion is invaluable to help sell the merits of a product line to his engineering colleagues. Meeting with engineers to review a product line is crucial in order to give credibility to the plan.

Engineers also provide valuable feedback regarding the viability of a roadmap in terms of schedule and feature implementation. This process is very much a balancing act between what is possible and what is convenient for engineering. Some engineering groups are bold, others timid. Marketers need to take caution not to promote a product line that is technically impossible to develop. At the same time, a healthy amount of risk is beneficial. Without any risk, a product has limited chance of becoming a true success.

COMPONENTS VERSUS SYSTEMS BUSINESS

In the early stages of product planning, marketers must decide if they plan to provide system solutions or a set of components for a system solution. For example, a player in the 802.11X wireless semiconductor market can either model itself after Conexant, or a company that just provides the radio section of the solution.

Conexant has a solid position in the 802.11x product segment and has developed a system-level solution necessary to bring a complete 802.11X wireless LAN product to market. This means that Conexant had developed the radio, MAC, and base band components, which utilize the company's complete software and end drivers.

In contrast, some companies may decide to focus on the radio portion of the solution. These companies may eventually be forced to develop a complete solution to defend their radio business.

Companies that develop the radio portion of a wireless solution do not have to develop the system solution or offer software drivers for its complete system solution. The job of developing the software and the system solution lies with the companies providing the base band and the MAC portion of the 802.11x products.

This is a strategic decision for a company that is only selling a piece of an entire system solution that could in the long run drastically reduce the company's chances of being a competitor or leader in the 802.11x market. However, if the company is in the business of developing RF components as well as high-end digital and mixed-signal products, the strategy may be valid if the company decides to enter the market later with a complete solution. By selling a piece of 802.11x technology today, the company gains essential knowledge regarding the dynamics of the market, which could help it greatly when it decides to launch its complete solution.

In any case, pursuing a system versus component strategy is a first step toward defining a roadmap. This decision has major ramifications on how resources are organized in the product development and execution phase.

INTEGRATION RACE

As mentioned before, a good marketer will always be mindful of the other companies that share space on a customer's printed circuit board (PCB). If a company chooses to provide just a portion of an entire solution, chip companies that aggressively integrate various components of a system solution onto their SoCs will eventually pose a serious threat.

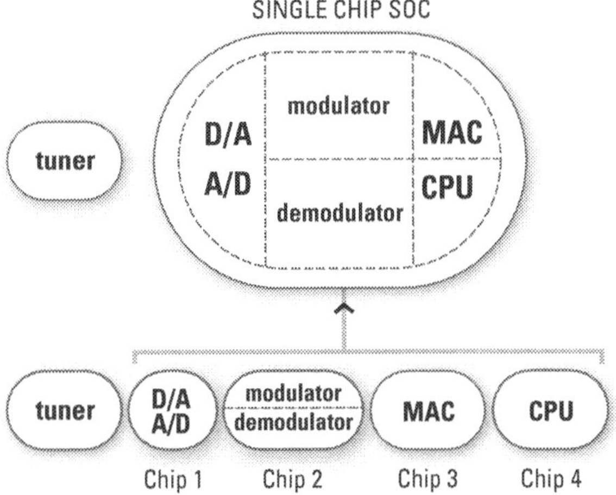

Brooktree Corporation's RAMDAC product line is a good example of the situation described above. Brooktree was a successful fabless company in the '90s that lost the integration race and was eventually bought by Rockwell Corporation. The Brooktree RAMDAC was an innovative product and one of the first to successfully use mixed-signal CMOS technology. Brooktree's technology cost-effectively enabled PCs to deliver high-end graphics to consumer PCs. RAMDACs were deployed at the end of a chain of components that included the graphics controller that companies like ATI produced. Unfortunately for Brooktree, graphics-controller companies started integrating RAMDAC functionality directly into their controllers. Brooktree caught on to this trend a little too late. An attempt to enter the graphics controller market failed, and Brooktree eventually sought refuge through a merger with Rockwell.

PC technology is dominated by the ever-growing strength of the main system processor and the system operating software. In retrospect, the RAMDACs never had a chance as a stand-alone product. Graphics controllers may someday fall prey to this trend as well. One might bet that the future of NVIDIA is not promising due to the increasing power of Intel's Pentium processors. However, one could argue that graphic-intensive applications such as gaming require continuous advancements in graphics technology. And as long as NVIDIA can continue delivering high-end graphics for the consumer PCs and game console market, it has a good future.

Many companies that successfully provide a single component for a system solution often underestimate how quickly competitors can integrate them out. Don't assume that a one-hit wonder will sustain itself indefinitely. The semiconductor industry is filled with examples of integration killing off product lines and companies. A marketer should always consider what functionalities make sense to integrate into future products from a technical and economic basis, hence making feature integration a key part of strategic product planning.

SOFTWARE ROADMAP FOR SOCS

If a product line consists of a system solution, versus component solutions, software will play a key role in making the product line successful. The cost of software development for a SoC is often many times the cost of the SoC development. In most cases, semiconductor companies (and in general the key executives) refuse to believe that they are in the software business as well as in the semiconductor business. This could be partly caused by semiconductor companies' trouble charging for their software.

However, SoC semiconductor marketers should understand that software is the enabler that sells semiconductor products, and that software creates the differentiation and the barrier to entry that is needed to protect a product line.

Marketers are learning to surrender the notion that just providing sample code for products is adequate for selling chips to OEMs. In most cases, OEMs are less inclined to handle software development around system solutions, especially for less complex systems. On the other hand, for more complex SoCs targeting closed-end solutions such as a proprietary set-top box, OEMs are prepared to take responsibility for software development. For markets that mature fast and are entering a high-volume commodity phase, such as the 802.11x wireless LAN market, a semiconductor marketer must have a complete software strategy that enables OEMs and ODMs to bring products to markets rapidly, often in a matter of weeks.

If a semiconductor company does not offer software for its system solution, then an OEM has to develop it. This may not be a problem as long as the OEM can derive value from the software development. However, as the value of the software decreases, the OEMs will be less inclined to invest in software development. Hence, OEMs demand that silicon systems vendors give them all the software they need to bring their products to market quickly.

A semiconductor marketer must determine what it takes to make products successful in the marketplace and decide how much software is required to support his products. A roadmap must clearly articulate a software strategy to both an internal organization and customers.

Often, a semiconductor company entering a new market has to develop and deliver a complete software stack in order to lower the switching cost associated with designing out incumbents' solutions. To do this, a semiconductor marketer must have a clear understanding of all the key features of an OEM's software requirements. This can be done if the OEM is willing to provide these requirements. Otherwise, a third-party organization must be available to validate and approve the software stack, or the software application, according to a standard specification. For example, if a SoC needs drivers that have to be approved by Microsoft, a company must submit the software for testing to Microsoft's Windows Hardware Quality Laboratory (WHQL) for compliance testing.

But things get a lot more complicated for products targeting proprietary systems. There a semiconductor company may be faced with the never-ending task of writing software products based on a set of specifications and testing environments that are not clearly defined or enforced.

In developing a software strategy for a product line, it is important to pay special attention to the increasing power of microprocessors and micro-controllers that are the key part of most end-system solutions.

Remember that integration can either be harmful or helpful. An example of this can be seen in the dial-up modem functionality of PCs. Until recently, dial-up modems were comprised of a chip set that included the data pump and a line side device. Today, V.90 modems are fully software based running on PCs, due to the increasing power of Intel's Pentium processor. V.90 modems were once selling for more than $50; today they are well below $10.

Another factor to consider when creating software requirements for an IC product roadmap is the cost of software development and how to account for it. As is discussed later, the industry has decades of experience in costing and making a profit selling chips. However, most semiconductor companies have difficulty determining the true value of their software development and how to charge for various products that utilize internally developed software. Often, software product development costs are under-estimated or not accounted for correctly especially in the production phase. This makes the pricing of the final chip difficult and in some cases results in a product line with negative contribution margin, even though on the surface the product line is shown to have positive margin. This can happen because the margin is typically calculated as sales minus the cost of goods sold, which is comprised of wafer cost, packaging, and test cost; the per-chip "cost" of the software is not captured.

In order to prevent this situation from occurring, it is best to monitor all activities associated with ongoing software development and support for systems products. If these steps are not taken, eventually the finance people will realize that the product line is not profitable, and that often means the end of a product line. In addition, it's necessary to understand software engineering resources needed to support a product line. Be wary of promising statements such as "All we have to do is create some drivers, and the customer can do the rest. So we only need a few software developers."

Software is a critical component of a system solution roadmap. A marketer must determine the importance that software will play to complete a product offering. Any software strategy should also ease the design-in process for customers. During product planning, marketers must determine the software investment necessary to develop a system solution and to support it.

In most cases, semiconductor customers do not like paying for software that is provided with ICs. In fact, as noted above, most IC companies develop software as a necessary step to design in their products. How value is extracted for software depends on many factors. If a complete set of low-level drivers is enough to help a customer design-in a product, then don't expect a customer to pay for the drivers separately. However, if a complete set of software solutions exists from third-party software vendors for a particular application, then a company may be able to charge for its internally developed software, as part of the chip price. The value of software will continue to increase as chip prices continue to fall, while new process technologies continue to make transistors effectively "free". Semiconductor companies adept at extracting value for their ever-increasing software investments will have strong competitive advantages.

ROADMAP FOR ESTABLISHED PRODUCT LINES

The marketer in charge of an existing product line, for which the company has spent years and many millions of dollars to develop and create a franchise, must maintain and expand an existing roadmap. Established product lines that enjoy significant market share demand constant vigilance to ensure that no holes exist in the product line with which a competitor can gain leverage. Market expansion may make sense for an established product line depending on product lifecycle or whether further territory expansion is justified from a ROI standpoint.

For a "maintain and defend" roadmap, the first place to keep tabs is on product costs. An incumbent must continuously lower the cost of manufacturing for an existing product in order to prevent any new competitors from entering its space as well as to maximize margins for that product line. In addition, planning should include a cost-reduction roadmap, which may include using new process technologies, if applicable, or new chip designs that result in lower die size, and/or lower chip costs from a packaging and/or test standpoint.

An existing product-line roadmap may also include product variants that serve multiple market segments. It is not uncommon for a competitor to develop a niche in an unrelated space and then using that position to attack a major market segment. Regardless of a company's position in a market segment, it is crucial that it does not alienate its customers by believing it is the only supplier in a particular segment. Arrogance is always detrimental to longevity.

Marketing an existing product line in a mature market segment is like operating a farm. A marketer must cultivate his products and at the same time defend them from hungry marauders intent on devouring his precious crop. New entrants have to compete on price or utilize windows of opportunity such as wrinkles in the technology landscape in order to enter an established market segment. The roadmap for an existing product line should shore up vulnerabilities in terms of cost and technology.

MARKET ENTRY CONSIDERATIONS

Developing a roadmap for a new product line is similar to starting a new company. Either the product establishes a new market for a new set of applications, or the product enters a market during a technology or channel distribution discontinuity. Either way, timing and product features must match customer needs almost exactly to allow for creation of a new market or entrance into an existing market.

illustration 14: success factors for established product

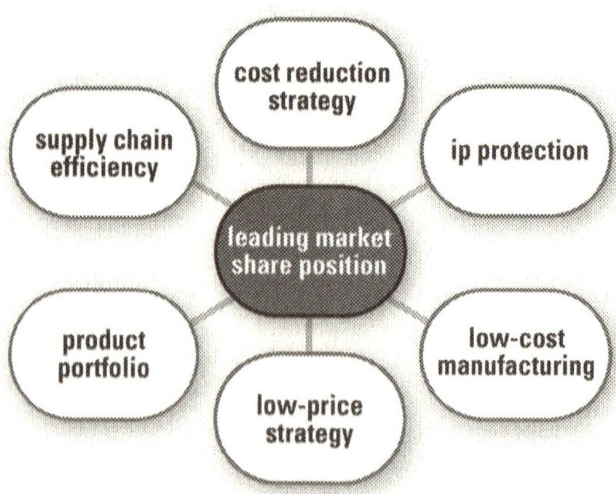

Technology discontinuities offer a good entry point for new companies to offer what the incumbent suppliers lack. We can once again look at the 802.11x LAN wireless market for examples of this. The first generation of wireless LAN products were mainly based on the original 802.11b standards. However, a couple of new suppliers started promoting 802.11g, which offers higher data throughput, to move the customers away from the 802.11b products. In addition, the 802.11g standard gels well with customers' perceptions that a higher data throughput product is better, even though most customers do not know whether they really need or can utilize the higher throughput of 802.11g.

Because of this situation, at that time any company with an 802.11g solution with the same cost point as the 802.11b had a definite advantage in the market. Clearly, any company considering a foray into the wireless LAN market with a generic 802.11b product found itself in some difficulty. The new 802.11b vendors found themselves with a "me too" product which could only compete in price and customer service, making it very difficult to win business against 802.11b incumbents.

To enter mature markets, a marketer must offer something beyond just the semiconductor product. For example, a marketer may decide to offer a software solution or a complete reference design that lowers the switching cost associated with customers' efforts to move to his chips. However, a target customer may be reluctant to move away from his software investment, especially if the customer's software team is already developing software for an incumbent's chips.

The marketing effort required to establish an entirely new market is also very challenging. Marketers must come up with ways to solve a business problem that customers want to solve, or create a demand for something that consumers desire. Larger companies generally have more resources to establish new markets. Companies like Intel have the money to do an effective job of creating demand for a new product line. Demand creation is not a simple task; it requires a tremendous amount of time and resources to change the cultural history of consumers.

We often see the PC OEMs, such as Dell, as a good gate or avenue for determining if a new technology has a chance of being successful in the marketplace. Dell, with its huge volume of PC shipments, is continuously trying to determine the true needs and wants of customers. Again, "need" refers to something for which customers are willing to pay to satisfy that need.

Companies such as Microsoft and Intel spend huge sums of marketing dollars to test-market and create market needs to push their future technologies. If you get a chance, visit the Microsoft and Intel booths at industry shows such as the Consumer Electronic Show (CES) in Las Vegas. You will see how a host of new products from both of these huge companies and their partners showcase where they want to take consumers. Do we really need to convert all of our VHS tapes to DVD format and save them forever in a digital format? The answer is that we really do not. However, Microsoft, Intel, and other consumer electronic firms such as Sony and Philips, benefit tremendously from this trend. Therefore, these large companies often engage in comprehensive marketing campaigns by turning consumers' desires into needs.

A small- to medium-size semiconductor company will not be able to change or create markets for its new products unless there is a powerful system or service company behind the effort. For example, let's assume there is a company developing a new product that helps telephone companies offer video services to homes. It first has to understand which suppliers currently offer networking gear to telephone companies. Then it has to make sure that marketing strategies of those companies match its own product strategy. Otherwise, creating demand will be a challenge.

Consider the networking space, where companies such as CISCO, Lucent, and Nortel have tremendous market power in creating demand for end-system solutions.

Even with their immense marketing power, these system companies can't sell their products if the service providers are not ready to roll out service to customers that utilize the latest networking gear. Service providers such as telephone companies, cable operators, and satellite service operators are constantly looking at new sources of revenue to grow their businesses. However, these companies also have to deal with the realities of the new capital markets and cost of capital associated with offering new services, and acquiring new customers. A chip company must understand these market dynamics and match their product offering to these dynamics.

BUILDING SUPPORT FOR ROADMAPS

Many semiconductor marketers fail because they do not take the time to build support within their own company for their product lines. It's not enough to have the most innovative and compelling roadmap and huge customer buy-in. Engineers and key managers inside the company must also buy into a roadmap, otherwise success of a product line could be limited.

Creating a roadmap based on market segmentation and positioning is just the first step in realizing revenue-generating product lines. The next step is to obtain the needed support from within the organization to convert the product roadmap from a slide presentation into actual products. In this sense, marketers perform "double duty" evangelizing not only to customers, but also to internal engineers and senior managers.

illustration 15: power of promotion

Marketers must defend their roadmaps within their company, by being able to respond to tough questions convincingly, based on their market knowledge. A marketer who does a thorough job analyzing a market and who gets a plenty of customer feedback, should have an easier job of selling his roadmap internally to the company.

The ability to evangelize a roadmap is especially important in companies with multiple product lines. Larger companies are often resource-constrained. Therefore, marketers need to compete for resources and the funding necessary to maintain and grow a business. Effective evangelizing also has the goal of attracting the best and brightest engineers to work on a product line.

If we take the view that an internal organization has many characteristics of an actual customer, then the process of creating support for a roadmap is one of marketing and selling a product roadmap. Marketers need to start by finding the key decision-makers and influencers who share a general view about the direction of target markets. For example, let's assume a marketer wants to develop a highly integrated multimedia chip. By doing some research, he could find out which of the top designers has the passion and the background for developing advanced multimedia chips.

Marketers must get to know the key designers. To a large extent, the world still turns on relationships. Developing key relationships will make the selling process easier. A champion in the design group is critical for developing momentum for a cause.

Having good engineering grassroots support is also useful in getting the buy-in of top executives in a company. Often top executives want to hear from the key designers about a proposed product line or a chip before giving their initial go-ahead to a marketing manager. Top executives know that without the support and willingness of key designers, success of a product or a product line will be jeopardized. In cases where top executives began their careers in engineering, the opinion of key designers take on even greater importance.

In addition to key designers, marketers must also seek the help of finance people to support their roadmap. The finance groups control the flow of funds to various product lines and must be convinced that there is a real chance a product line could be successful in the marketplace. A finance group often acts like a gatekeeper and is usually free from a technology bias. This objectivity is needed as a form of checks and balances. A good finance group will help tone down over-exuberance in product planning by monitoring proposed development costs and financial return.

ROLE OF CONSULTANTS IN DEVELOPING PRODUCT ROADMAPS

Companies can sometimes spend too much time developing a roadmap. This is mainly due to disagreements among various camps within the company coupled with a lack of true market data about customer needs and wants. Typically, marketing groups make this situation worse by failing to gain popular support for their roadmaps, which can be due to inadequate preparation or market knowledge. It is not uncommon during these periods that engineering effectively becomes the marketing group and does what it thinks the market needs and wants.

To correct this situation, semiconductor companies often turn to consulting firms to help them develop product-line strategies. The services of the consulting firms are usually obtained at the direct request of a company's CEO.

CEOs have the ultimate task of setting high-level corporate strategies that best utilize the assets of the company, given a set of market conditions. Consultants not only straighten out warring engineering and marketing groups, but also help develop a top-down consistent strategy throughout the whole company. This is especially important in companies where top-level strategies have trouble trickling down to the working level staff. Companies lacking adequate management communication frequently develop product-line strategies that are not in line with the stated goals of the CEO.

The role of consultants in setting product roadmaps can be very effective, but this activity requires the full support of marketing managers.

Consultants need to obtain an in-depth understanding of product-line managers' views on various markets to help develop effective product roadmaps. However, many product-line managers resent the role of consultants and think that they do not need any help from outsiders to create an effective roadmap. For a consulting relationship to work, managers should put aside their prejudices and work with the outsiders in order for the relationship to be successful. At the same time, managers at all levels must constantly challenge the perceptions and conclusions of the consultants. This is essential to ensure that the consultants are putting the requisite research and analysis into their strategy development.

Consultants can provide an unbiased marketing strategy. This is important because marketing people often lose sight of the realities of their market and their product lines amid the stresses and tensions of the job. It is not uncommon for marketers to go into a denial phase when market dynamics change for a particular product line. Denial is very dangerous, especially if the window for corrective action is narrow. Consultants can remedy these situations by anticipating the "train wreck" and advising corrective action early.

SUMMARY

Roadmap development is a critical component of overall strategy for a semiconductor company. The roadmap is an important planning tool that displays a marketing strategy and how a company plans to dominate a market segment. Internally, it is a useful planning tool for engineering, finance, and marketing. Externally, it is a promotional tool used to convince customers to buy into a company's vision and product lines.

Roadmaps are also used to judge the worth of marketing people. Senior managers are constantly evaluating the ideas presented in roadmaps. They judge marketing on their ability to articulate a coherent and believable plan. They look for someone who can speak confidently about a market and about where that market is going.

The product roadmap is the key to setting a product line's short- and long-term strategies. For success, the roadmap should be rooted in the marketing fundamentals discussed in chapter 1. The roadmap is a result of market analysis. It also illustrates the targeting of market segments over time. As an expression of vision, the roadmap also becomes an important promotional tool for customers. Done correctly the roadmap is critical component to product line success.

developing the MRD >

The roadmap is a general "plan of attack" over a specified timeframe for a product line. While the products showcased in the roadmap may include high-level feature descriptions, these descriptions are not detailed enough to start product development. Marketing communicates the details of a specific product on the roadmap through a Marketing Requirements Document (MRD).

The MRD is a document that presents background information about the market, competitors, product features, and the required product development schedule. Unlike a roadmap that lists the evolution of a product line, the MRD is specific to a product. The MRD should be written any time marketing is requesting engineering and/or the business unit to invest in a significant development. While most companies use MRDs strictly to introduce new semiconductor products, MRDs can also justify significant software or systems work on existing products.

At a minimum, the MRD should contain the following:

- General Product Description
- Market Overview
- Market Segmentation Analysis
- Product Definitions

MARKETING REQUIREMENTS DOCUMENT STAKEHOLDERS

MRDs represent the first serious statement by marketing on a market need for a specific product. It is also a document that guides engineering. In well-managed companies the MRD evolves into a systems specification that later evolves into a detailed technical datasheet of the product.

A product MRD is usually distributed to engineers, marketing, and top-level executives, those who have some stake in the product line. In most cases, it takes one to three months to develop a comprehensive MRD. All the key decision makers should review the MRD and buy into it.

Engineers are usually very receptive to documented marketing ideas (they figure that marketing must be truly excited about a prospect if they bothered to write about it). Also, engineers are used to documentation as a means of communication. Engineers are usually excited to get MRDs as the document represents a new design challenge; it offers engineers the opportunity to start their creative problem solving skills.

Marketing must also "shop" the MRD with key managers. Usually senior management decides what new programs to fund. In a start-up this may be the CEO; in a larger company the decision-maker is typically the general manager in charge of both marketing and engineering. In some companies, there are detailed product development processes in place that must be followed to get new products approved. In other companies, the product approval method might be free flowing.

illustration 16: marketing requirements document stakeholders

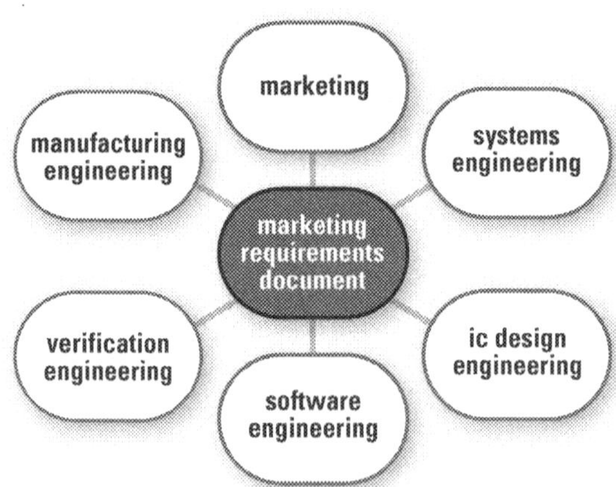

Marketers must spend adequate time writing MRDs. The content of this document forces marketers to analyze the needs of a target market as well as the competitive environment of that market. It also forces marketing groups to justify features required for a specific market. A problem typical in semiconductor companies is feature creep. Feature creep is usually a result of either engineering or marketing asking for new or an ever-larger set of features to be designed into a product.

The reasons for feature creep can be numerous. One reason may be lack of proper market segmentation and targeting. If a marketer is not sure of a target market, he will hedge his bets by adding numerous features to a product definition to make it more generic. If market analysis is done thoroughly and markets are truly understood and segmented in accordance with logical analysis, an MRD should be a solid document. Good analysis and a well-written MRD should limit the feature creep phenomenon. Feature creep should not be necessary even when market requirements change since a good marketer should anticipate those changes.

By the way, engineering can also be the source of feature creep. It is common for engineering groups to unilaterally decide to put in features they feel are important for whatever reason. Sometimes there are sound technical and marketing reasons for these additions – sometimes not. It can also happen that a company's prominent customer co-ops the company's engineering group. It is not uncommon in such a scenario for customers to dictate new requirements that bypass any marketing or management scrutiny. Once again, these additions may or may not be appropriate. A well-written and actively used MRD will help regulate this type of feature creep as well.

Finally an MRD should be considered a living document that is maintained and updated to respond to unforeseen shifts in market conditions or changes in resource allocations. When it is updated, all stakeholders in engineering, marketing, and management must be aware of the updates and changes. An MRD should never be written only to collect dust on some bookshelf. Unfortunately, due to the urgency of getting most products started, most companies do not pay the necessary attention to the development and maintenance of MRDs. This can lead to conflict among the team members on what a product is, should be, and how it should be developed. By using an MRD as a baseline, the document can help avoid internal conflicts by forcing management, engineering and marketing to focus on products justified by market analysis.

ESTABLISHING BUSINESS/GATE REVIEWS

By the time an MRD is completed, a lot of marketing analysis and work has gone into the development of a product line concept. This means, marketing has identified a market, targeted a promising market segment that matches the strengths and strategic positioning of the company, built a product roadmap that provides "plan of attack" for the target market, shopped the roadmap to customers and internal stakeholders, and completed the MRD for the product.

Before any work can begin by engineering, a marketer must justify the investment in a business review to the key managers of his company or business unit. The business review is similar to how a start up company might pitch a business plan to a venture capitalist. The basic dynamics are the same in that someone with a marketing idea is trying to convince someone with money to invest in his ideas. In the case of a semiconductor company, marketing must convince management to spend company resources on a new product development.

Most semiconductor companies do a business review or "gate review" to vet new product ideas. The review includes both a business justification as well as a technical feasibility study. The advantage of a review process is that it can involve all the stakeholders (for example: design, test, product, system engineering, management and marketing) in a product line. Given that the development of a semiconductor product has many stakeholders, an iterative review process is usually a requirement. Some companies require lots of extremely detailed and lengthy reviews while others do not. The need for iterative reviews has gained in importance as semiconductor products grow in complexity from a chip, systems, and software standpoint. Processes vary widely given the type of products and management styles of a given semiconductor company.

The following discussion on business planning is a simple overview of a very involved process, the complete details of which are outside the scope of this text. This section will provide the reader a quick overview of business planning for a semiconductor product. By the way, marketing is usually the initiator and driver of the business review process. But a product idea can come from anywhere in the company. However, even if the original idea comes from engineering, marketing will usually take the responsibility to drive the market analysis and business review process.

TOPICS DISCUSSED IN BUSINESS REVIEWS

The business review is usually a series of meetings between engineering, management and marketing. Usually marketing is presenting in these reviews. Regardless of a specific procedure used by a company, these reviews at a minimum discuss and confirm the following points:

1. The definition and feature set of the proposed product and its contribution and strategic value to a product line roadmap strategy.

2. The Total Available Market (TAM) of the targeted market segment in terms of units and dollar value, assuring that the TAM is compelling and appropriate to the company's strategy.

3. The Serviceable Available Market (SAM) of the targeted market segment in terms of units and dollar value, assuring that the SAM is compelling and appropriate to the company's strategy. If a product is well defined for a granular market segment it is possible that SAM will equal TAM. However, most semiconductor companies use SAM and TAM if they want to highlight market segmentation analysis in their business plans or if there is strong competitive incumbency in the market segment. If one product can serve multiple market segments then it is useful to use SAM and TAM to highlight how the market numbers change given different features of the final product.

4. A detailed discussion of the competitive environment. If there is an incumbent in the target market, it is necessary to provide a complete analysis of the competitions' product features benefits and weaknesses.

5. The product development feasibility within the market window given the technological capabilities and resources available. This analysis must include analysis for the chip development, and any software and hardware support required to make the chip work.

6. Total projected investment. This total includes all projected cost to bring a product to market including design, product, test, software, hardware, and marketing costs.

7. The business and technical risks assessment, given the level of investment exposure by the company.

8. The required final product cost target.

9. The projected return on projected investment dollars. This needs to be adequate and in line with the company's expectations and strategy.

RETURN ON INVESTMENT (ROI) ANALYSIS

In most cases, managers focus the most on the last line item in the above list. But, it may not be the only factor that determines whether a company decides a course of action. Ideally, the ROI is a large number. It makes sense that managers want to maximize the money they make from their investments. Making money is usually the reason why a company exists. However, the final "yea or nay" on whether to start a product may also include strategic considerations. For example, if a marketer is going after an existing market, it is quite possible that the first product in a roadmap will not have a great ROI projection. The company may decide to accept a low or negative ROI in order to "get into the game" or "stay in the game". If that is the case, then it must be justified by the strategy outlined in the roadmap because the company will expect revenues and profit in subsequent product developments.

ROI is not a straightforward calculation. Given that most semiconductor projects take a considerable amount of time in terms of development and product roll out, marketers must use time based ROI tools such as Net Present Value (NPV) and Internal Rate of Return (IRR) analysis to compare short term investment costs against longer term revenue and margin projections. These mathematical tools are well documented in basic business texts as well as on the Internet.

During business planning marketing is also tasked to project the profit generated by the sale of a chip. In order to calculate this, marketing must estimate the average sales price of a product. (Establishing a price for a product is treated in detail later on in this book.) In order to arrive at a profit margin, the cost of the chip must be projected.

Cost analysis can be very complicated. Usually, sophisticated "cost models" are employed to calculate costs. But basically it is derived from the wafer cost for a given process technology, number of dies per wafer, yield estimations, assembly and packaging costs, and probe and final test costs. In addition, any royalties that need to go to third party technology contributors are added to the final per chip costs.

Once a chip cost is determined, gross margin dollar can be calculated as the ASP minus chip cost. Gross margin percentage is also calculated using the following equation:

GROSS MARGIN PERCENTAGE = (ASP - CHIP COST) / ASP

Normally, the finance people in a semiconductor company will provide any additional percentage cost that needs to be added to per chip costs to derive standard margins. These additional costs are added across all products and represent the cost of doing business such as cost of sales and operations. Generally marketing focuses on gross margins since the overhead costs are out of marketing control. Typically a semiconductor company can have anywhere from five to 20% added in overhead costs per chip. Whether a projected gross margin is acceptable or not depends on the financial strategy and objectives of a company.

Once per chip cost and ASP projections are available, marketing can apply this to a SAM estimation to calculate total margin and revenues projected over a given market window. The market window depends on the type of product developed and the characteristics of a target market segment, mainly the revenue-generating life of the product. ROI is then a comparison of the NPV of projected earning versus total investment.

The percent IRR of the initial investment is also useful to compare with the current rate of return of typical financial investments such as stocks and bonds. Fundamentally, a marketer should always ask himself whether an investment in a product line is more or less lucrative than simply investing in stocks and bonds. This way of looking at things should not be limited to marketing. Looking at product investment as any other investment is critical for the development of a marketing mentality throughout a company. Everyone in a semiconductor company must realize that the purpose of their employment is to make money for the company stakeholders –such as company employees and stockholders.

One might ask "why invest in developing new semiconductor products at all, why not just put the money in stocks." Once again the person with the right marketing mentality can easily answer this question. The reason we invest in semiconductor products is because they can be extremely lucrative. Typically, a company will not embark on a new semiconductor product unless the ROI is at least a factor of 5 or 10 times the value of initial investment (unless there are some strategic factors). If the product is successful, the ROI can be a much larger multiple.

MRD PROMOTION

For many people outside of marketing, the concept of marketers writing MRDs and preparing slides for business reviews is probably not surprising. It sounds like activities marketing should do. But the truth is, marketing is also about selling a vision and leading a group toward that vision. This selling process starts with the evangelization of a road-map and culminates with an MRD and business review. In fact, the MRD and business review are just as much about internal promotion as they are about marketing analysis. A marketer must persuade his company to spend valuable resources on a plan.

For this reason, skillful marketers are very good at developing key relationships "behind the scenes". Just like the selling of a roadmap, it is also important to keep up relationship building with the MRD and business review. For example, systems engineers and key designers should be involved early on in the development of an MRD. In fact, in some companies, systems engineers work closely with marketing to write the MRD. Systems engineers are in a great position to review high-level features in an MRD and turn it into a more detail feature requirement. Systems engineers can also ensure that a feature requirement is properly explained. For example, your product may need a specific I/O port. Systems engineering can help marketing determine more details like the speed requirements for the I/O port and whether it has to be a serial or parallel port. More importantly, systems engineers may be the first group of engineers to start considering how a product can be developed.

Marketing typically takes key system and design engineers to visit important customers for roadmap MRD discussions. If the product targets a market segment that is dominated by a few customers, it is prudent to review the detailed features of a product proposal during face-to-face customer meetings. These customer meetings also serve another purpose. It is an event where marketing can build support for the product effort. Face-to-face meetings between customers and key engineers are a great way for engineers to learn firsthand what customers' requirements are. In addition, these meetings are especially important for engineers who have not bought into a proposed product idea. Visiting customers together can also help develop a sense of ownership among marketers and engineers.

Another reason marketers must promote their product ideas internally is to ensure that the top engineers are enticed to work on their product lines. Having the best engineers work on a product line will help ensure success especially from a development schedule standpoint. Without having smart people working on your product line, you will end up with mediocre products that never make it to market in time.

SUPPORT OF SOFTWARE DEVELOPERS

Marketers must also promote their products to key software engineers. Software is becoming a more important component of most ICs. Too often, semiconductor marketers lack skills in understanding software or even how to market software. Not only should marketers sell their product concepts to software engineers, they should also develop relationships with them in order to learn as much as possible about software development and productization.

Semiconductor companies usually struggle with developing software. This starts at the driver level and increases as higher layers are added to a software package. In some cases, there is no true software engineering discipline. Drivers are often developed and maintained by hardware engineers who lack sufficient software development experience. The situation is made worse when marketing people underestimate the software development efforts necessary to bring a semiconductor system solution to market.

Semiconductor companies also pay more attention to hardware groups than software groups. If software is a necessary component to a product line, it should be included in the roadmap as well as the MRD. A software roadmap shows what level of software is necessary to make the product successful in the market.

Unlike the effort required to bring a chip to market, software development is an ongoing effort. As the saying goes, "Software is never finished." It must be maintained, upgraded, and improved. In most semiconductor companies the task of sustaining product and supporting ongoing manufacturing issues does not reside with the original chip development groups. Product and test engineers are the people who take over support of products after the initial development phase. However, product and test engineers are primarily focused on hardware, not software, issues.

A good marketer should never underestimate the complexity associated with developing and maintaining software. At a minimum, underestimation will run the risk of losing the support of a software team working on a product line. Underestimating software development during the

CHAPTER 5

business review and project planning phase can also lead to resource understaffing, which can lead to trouble for a product line.

SUPPORT OF PRODUCT AND TEST ENGINEERS

Marketers must also work on building solid relationships with product and test engineers. Product and test engineers are responsible for maintaining the yield, quality, and reliability of products during their entire lifecycle. Product and test engineers should be consulted during the development of an MRD and during business reviews. There is more to shipping high-quality semiconductor products than just creating a design and then throwing it over to manufacturing. Product and test groups play a key role in the success of any product development effort.

Most often marketers and product engineers clash on certain issues. Typically, marketing wants quality products available as soon as possible. They also tend to overlook the time and resources needed to achieve this. Product engineers, on the other hand, are often reluctant to move through product characterization and qualification at a speed that could negatively impact product yield, quality, and reliability.

Taking time to develop relationships with product engineering teams will yield great dividends during the initial stages of product development as well as once products move to full production status. Working with product and test engineering in the early stages of writing an MRD and during business reviews is a great way to get them excited to work on the product line. They will also help move the business review discussions along if they have been fully briefed prior to official meetings with management. In addition, not only will a good relationship with product and test engineers bolster their interest in working on a new product, it will also help these engineers make good decisions on how to bring a new device to production faster. They will be more willing to look for ways to expedite their processes since positive interaction with marketing makes them more informed about how a product will be used in the end applications. This will help them focus on verifying those parameters and features that are truly important for the operation of the new product in system applications.

A marketer should take the time to present his product roadmap, MRD to the product and test groups. The marketer should communicate the fact that he has done an in depth analysis and that there are exciting markets out there to profit from. During these presentations, the marketer can show engineers how their work directly contributes to the success of a product line. Particularly important is to show how the cost of products and product yields contribute to the bottom line. (Everyone must understand that revenue generated for a product translates into such things as salaries, jobs, medical benefits, and all the other "goodies" associated with stable employment.) Also, marketers should make it clear how product delays can contribute to losing customers, market share, and revenues. If product and test engineers feel part of the team, they do everything possible to ship high-quality, high-performance products to market as fast as humanly possible.

SUPPORT OF HARDWARE ENGINEERS

Hardware engineers are good people to work with in the early stages of MRD development. System engineers are tasked with taking the finished ICs and making them work in actual systems, a necessary step for product evaluation. Similarly, package designers responsible for chip package design and substrate development are also critical stakeholders in MRD development. By working with them early, they will quickly point out the best approaches and potential pitfalls of proposed feature sets and cost targets.

Winning over engineering must start early with the roadmap, the MRD as well as the development of the business review. A marketer who works in a box will surely fail. A marketer who involves engineering only for technical guidance may or may not succeed. A marketer who takes the time to really sell his vision to engineering will have the best chance of success in the business review process. Management will have a harder time saying no to a new product idea if marketing and engineering show passion and enthusiasm for a new opportunity.

SUMMARY

The roadmap, MRD, and business review culminate in a decision, usually by a top manager of a business unit to either embark or not on a new product proposal. The decision to embark on a new proposal does not mean a marketer's marketing work is over. Marketers must constantly analyze markets and keep an eye on product development and schedules. Internal product evangelization should also be practiced to keep everyone on schedule and excited about a product line.

The true test of a marketer's analysis of a market opportunity is yet to come. It is during the long development phase that the marketer's assumptions of where the market is going will be tested. It is every marketer's fear that a competitor will rise out of nowhere with a new product and take advantage of a target market segment. It can be a stressful time that requires for all member of the team to execute quickly. Marketing is often in charge of program management, especially those involved in product marketing. The next chapter discusses how effective program management can ensure product development success.

With management approval of the business review, the time has come to turn marketing dreams into reality. After the strategic process of identifying a market, segmenting and targeting the market, developing a roadmap, and creating an MRD, marketers must focus on product execution and roll out — more tactical activities. Engineering groups control product development. However, product marketing is either officially or de facto in charge of program management. Whatever the situation, marketing should be very active in supporting program execution.

PRODUCT DEVELOPMENT PHASES

Product development cycles are often twice as long and twice as expensive as originally planned. A semiconductor marketer must understand the product development process and put a system in place that helps track a program and ensures that multiple development activities are working in harmony and along a master schedule. Throwing the job completely over to engineering and expecting products to be developed on time and with the necessary features is wishful thinking. Marketing people need to set up processes to ensure that development projects get done within a time frame, a planned budget, and according to specifications required by a target market.

The end goal is a successful product launch. Introducing a product to market not only means that the product is complete, but also that the support and promotional programs, pricing, and documentation are complete and ready to go. Program management is critical from the beginning of product development to ensure the necessary elements required for product launch are also ready for market introduction. This is true for any semiconductor product; it is absolutely essential for highly integrated SoCs.

The initial basis of any program planning must be an MRD. Before development teams start serious design work, the MRD must be complete.

All parties must agree on the product's features, development cost, software requirements, die size estimates, sample schedule, application goals and production ramp schedule.

Marketing should also be thoroughly convinced that the MRD is correct on stated product requirements. There should not be a problem as long as marketing has done a thorough job on the MRD. If marketing vacillates too much on product requirements or feature priorities, engineering and management will question the credibility of the project — and rightly so. It's true that markets are dynamic and requirements may change during product development. However, to some degree marketing is being paid to anticipate the direction of a given target market, and develop products that take advantage of market dynamics.

It usually takes 12 to 18 months to develop a new semiconductor part. The cost of developing a new product can easily approach $10–20 million. Therefore, defining the wrong product can be an expensive mistake. In fact, it is even more expensive than just the sunk R&D costs — since future revenue and valuable time will also be lost during a period the company could have been doing something productive.

illustration 17: product development phases

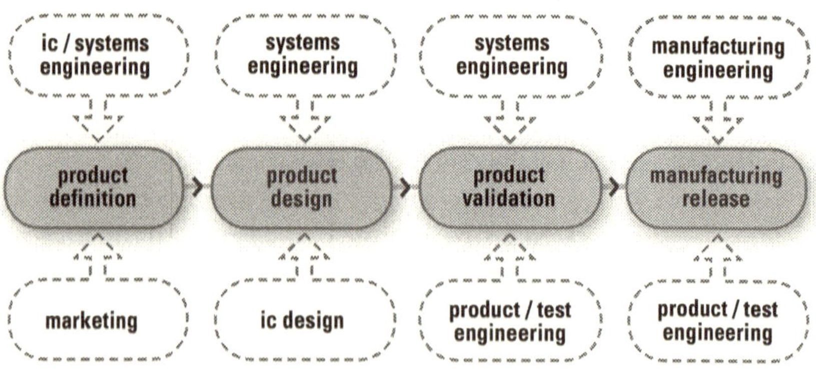

Semiconductor companies generally use multi-functional teams to bring a product from the concept phase to full volume production phase. In order to release a product to volume production, full cooperation of the design team as well as the product, test, and quality-engineering teams is a must.

DESIGN FOR MANUFACTURABILITY

As program managers, marketers must promote a "volume production mentality" to help induce development teams to design for quality, manufacturability, cost, and product support. It is best to avoid a "product sample/demo mentality". The team should be concerned about the quality, manufacturability and cost of the product from the early stages of the design phase. Simply testing products during manufacturing cannot optimize quality; it must be designed into the product. Manufacturing yield can be improved by ensuring that a product's design optimizes yields by designing for changes in process variations, operating voltages, and temperatures. Manufacturing test efficiencies can be improved by integrating Design For-Test (DFT) strategies. Cost certainly can be minimized as well by optimizing die size, package requirements and testing – all affected by the initial design. Finally, product support in application is also a function of design. For example, the efficiency of a Printed Circuit Board (PCB) layout is affected by chip pin assignments. By designing for volume production, design teams will work towards ensuring a quality product that is optimized for manufacturability, cost and support.

In addition to ensuring that multifaceted development teams are working in concert toward full volume production, marketing must also communicate the urgency of completing a design on time and on budget. Different managers use different techniques to communicate to their development teams the urgency of bringing new products to market. Some managers threaten the team and manage by fear, while others use more positive motivational methods.

We have not seen management-by-fear work in the long term. This method might work during extreme industry downturns where people are afraid of losing their jobs, but most often, it ends up breaking any genuine spirit of teamwork. Good marketing managers articulate a vision of success to their team members. It is important to convey market observations frequently and clearly to the team. It also helps to have an external enemy such as a competitor for your team members, to foster a feeling of camaraderie.

Marketing should not show much patience towards those who are destructive to the mission of the team. People who constantly emphasize the impossibility of overcoming challenges, or harp on the negatives, should be counseled or released from the team to prevent them from demoralizing others.

PROGRAM REVIEWS OBJECTIVES

To improve the success rate of your product development, marketing should insist on frequent program reviews to measure and control the progress of product development teams. In addition to large program reviews, product development teams must meet weekly to discuss issues and assign tasks to team members. Attendance at the weekly team meetings must be mandatory, and team managers must publish minutes of the weekly meetings within 24 hours of meetings. Team meeting should not be a soapbox for ranting, chatting, or complaining. These gatherings are a place for members to discuss their issues and problems and seek the help of the team to devise solutions.

Marketing together with key engineering managers should also schedule monthly meetings with upper management to update them on the status of a program. In these meetings, resource problems and other high level issues can be raised, where management's help is required for resolution. The minutes or notes from these meetings as well as program reviews and weekly team meetings should be posted on internal websites. This is especially important for teams that are separated geographically.

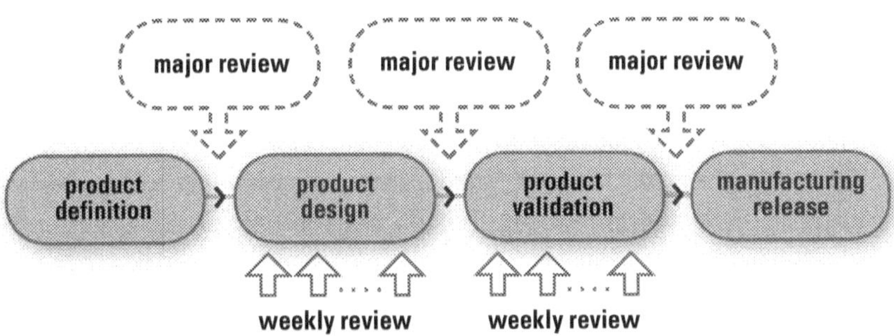

Marketing or program managers must also make sure that all action items from various team meetings and monthly review meetings are recorded and addressed. This brings a sense of discipline and accountability to team members and gives management the ability to quickly see the major issues the development teams are facing. Program managers should consider utilizing various web-based tools to enhance the communications of team members. In most cases, lack of communication is the primary reason for product delays and cost overruns.

In case of any delay for any part of the program, the goal of the responsible team members should be to analyze the root causes of the delays and communicate them to other team members quickly. If the team cannot solve the issues promptly, they must raise the issues to the upper management. Program managers should not be shy about asking upper management for help; they are there to manage priorities and communicate the overall directions of the company to everyone. It's a good idea to ask top management to drop by weekly team meetings to give the team members their support and thank them for their dedication and hard work. If a team is not performing, upper management must visit the team to suggest ways the team could improve its performance.

Remember: Hope is not a strategy. Hope will not fix product delays or the poor performance of any development team. Frequent meetings with development teams are a way to ensure that a program is on track. Marketing must be involved in these meetings even if a company hires separate program managers. Marketing must never sit back and assume that engineering will complete exactly the product required on time and on budget. Marketing ultimately owns the business, and must be proactive in ensuring that business will be successful. Marketing must ensure a program is on track and also ensure that people are kept excited about the possibilities of success. Marketing must communicate clearly to the team why they need to work hard and deliver products to the marketplace on time and on budget. This success is only possible through a true team effort.

As semiconductor products gain in complexity, more companies are hiring program managers as full time positions. Good program managers must have the right training to manage large and complex programs that need the participation of various groups within a company. However, these program managers are often placed in a functional team such as marketing, product engineering, or design engineering. In each case, these program managers end up having a bias toward the needs of their functional area. In addition, program managers who do not have the full support, and do not possess full authority, of higher-level management often fail at managing product development teams effectively.

A company must choose program managers who have both the understanding and appreciation for both engineering challenges and the business issues and urgencies. Program manager effectiveness will depend on their having the authority and the time they need to manage a product development team. If a program manager is placed to work in an engineering group, it is a good idea for marketing to work closely with the program manager to ensure that product launch requirements are tied into the overall development program effort.

LOW-COST PRODUCT STRATEGY

Semiconductor prices are usually under pressure due to the entrance of new competitors to a market segment, and the lowering of manufacturing costs in general. This means that, during a product development, a marketer must pay attention to how much it can sell a product for not only in the first months of its introduction, but also years after its introduction.

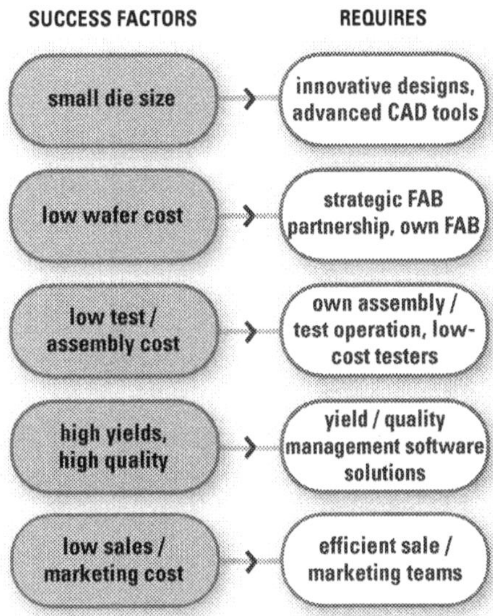

Therefore, a company's operations department must make sure they select a wafer fabrication supplier that can provide competitive wafer pricing as a product moves to its commodity phase. Marketers have every right to challenge operations on its vendor selection. They must also build their roadmaps to take account of the price erosion phenomenon.

Operations should establish strategic partnerships with key foundries. Without this, obtaining good wafer pricing can be a challenge. Imagine the difficulty of paying as much as 20–40% higher for wafers than your main competitors. To compensate for that wafer cost difference, designers must be able to develop devices that are 20–40% smaller than the competition, or they must find more value-added features to incorporate into products to capture higher prices. In any case, having poor wafer costs puts a product at strategic disadvantage with competitors.

In some cases, semiconductor companies make large sums of strategic investment dollars in wafer foundries, both at home and abroad, to make sure they get lower wafer pricing and higher production allocations. Setting high-level foundry relations should be one of the primary jobs of the CEO and high-level operations staff. It is marketing's job to communicate the realities of the marketplace to senior management.

Once operations have selected a strategic partnership with a wafer supplier, the design team should be ready to use the selected process technology. It does not help to get lower wafer pricing from a supplier if the designers are not ready to design products in that process. While marketing is not typically in control of either operations or engineering, it still should ask obvious questions. In some cases, there is a real breakdown between operations and design groups. Designers are always pushing to use stable and leading edge wafer foundries for their products; operations people are often chasing lower-cost wafer foundries. Marketing should keep vigilant to the possibility that miscommunications between operations and engineering could destroy a product line's chances for success.

In addition to wafer prices, marketing should monitor final product yields and test costs. Marketing must work with product and engineering staff to put controls in place so that proper actions are taken if yields drop below originally planned levels. Marketing must also ensure that the finance group uses the latest yield and manufacturing costs to calculate final cost of products. In some cases, the finance group may use outdated numbers that make your product appear to cost more to produce. Marketers need the true cost in order to make good business decisions.

PRODUCT CANCELLATION CRITERIA

One of the hardest things for a marketing manager to do is to kill a product during its development phase. This is primarily due to the fact that once engineers start a project, it's difficult to move them away from it and quickly re-deploy them. It's also to some extent due to the wishful thinking of senior management that the team can finally deliver and get the product out to market on time and within budget. In addition, senior managers may ignore the reality of the marketplace and the competitive landscape of the market they are targeting. Often a company does not want to admit its mistakes and to recognize the risks associated with continuing the project.

The semiconductor industry is highly competitive and unpredictable. It's extremely difficult to define and deliver hit products to market on time and on budget with a 100% success rate. Most companies fail to even have 50% product-success rate. Companies must seriously consider killing products that are in the development phase if those products are late or cost too much to produce.

It is especially hard for a marketer to kill a product; after all, a marketing person is normally the chief evangelizer for their products. However, sometimes business logic must dictate over sentimentalities. To ensure this, marketers must frequently measure the progress being made on a program and compare this with market realities. If a program slips too far, or if market realities drastically change to make a product irrelevant, marketers should be prepared to communicate this to their senior management. Marketers should strive not to launch inferior products to highly competitive market segments.

The decision on whether or not to kill a product should also be based on the strategic importance of that product to the company. For example, even if a product may not deliver any revenue to a product line, it might serve as an important building block that could be used in future products. Therefore, marketing may decide to finish the product development and characterize the part, but hold off introducing the product to market.

It's hard for anyone to admit they cannot deliver a product on time and on budget. We all want to be associated with winning projects and teams. However, if a product development has gone off track, and is beyond repair or saving, marketing must recommend canceling the product development. Marketing should cancel products that will not contribute significantly to revenue goals or have little strategic importance value due to changing market dynamics. It may not be a popular decision, but it is probably the right decision.

SUMMARY

Coaxing a product to reality is one of the most difficult aspects of a marketer's job, but it can also be one of the most rewarding experiences. The launch of a new product is a realization of an idea.

This chapter has presented some of the major points for ensuring product success. This success depends on making sure product requirements are complete and correct and that multi-faceted development teams are working towards the goal of launching products into high volume production. Marketers must either take the position of program managers or work closely with the program managers assigned to product development teams. In addition, marketers must continue to evangelize success of products in the market and update all team members on market developments.

Marketers should also work closely with operations and product teams to ensure that a product will enjoy the lowest possible cost once introduced into the market place. Finally, in its constant vigilance, marketers must also keep their eyes open to the possibility that the products under development may lose their relevance. In these cases, marketers should be prepared to advise upper management to discontinue funding of troubled projects.

Once a product is launched, the focus for the product line changes from an internal to an external perspective. The issues change to product roll out and promotion — the topics of the next chapter.

product roll out and promotion >

Months before a product is ready to sample with customers, marketing develops a product roll out plan. The product roll out plan is a checklist of specific actions necessary to complete in order to have a smooth introduction of a product into the market. This checklist is normally comprised of elements that are required to support a product in the market, as well as a list of promotional activities.

PRODUCT ROLL OUT CHECK LIST

The specific checklist of actions is variable depending on the product, the market segment, and the types of target customers. A typical checklist may include the following:

SUPPORT RELATED

* Datasheet
* Application Notes
* Software Documentation
* Supporting Performance Data
* Bill of Materials Estimates
* Evaluation Boards for hardware verification
* Evaluation Boards for software verification
* Reference design/Manufacturing Kit (if appropriate)
* Demonstration Platform
* Product Pricing
* Sales/Field Application Engineering (FAE) Training Materials
* Sales/FAE Training Schedule
* License Agreements
* Product set up in Order Management System

* Product Bulletin
* Product Presentation
* Press Release/Media Advisory
* Press Tour Planning
* Key Customer Visits
* Key Industry Show Schedule
* Print and Media Marketing Campaign
* Web-based Marketing Campaign

PRODUCT SUPPORT PLANNING

The ultimate support list necessary for a product launch will vary widely depending on the product and the characteristics of a target market segment. That said, it is marketing's role to clearly define what exactly is needed to launch a product into a market.

The fundamental question should always be, "what do my customers need in order to quickly design in my product." Note also that sources of sustainable competitive advantage can be found in the way a company supports its customers. Marketers should always consider new methods of support that provides an edge over the competition.

Typically, customers need documentation to understand how a product works. A datasheet, application notes, reference designs, and software documentation, are normally required for most semiconductor products. In some companies marketing is directly in charge of documentation. Other companies employ technical writers who work with engineering to develop professional grade documentation. Still other companies may rely on a combination of marketing, engineering, and marketing communication groups to produce product documentation. In any case, marketing must ensure that documentation is getting done in time for launching a product into market.

Sometimes a market may require that a semiconductor company at a minimum provide an evaluation board to help with customer chip selection and evaluation. This depends on the product and the needs of target customers. Marketing must decide what level of application support is necessary to win business. If a customer is extremely savvy, or if a product is very easy to use, an evaluation board or demonstration platform may be sufficient. In some cases, the only requirement for a board may be for internal testing. On the other hand, a marketer may be faced with an extremely complicated product and a set of unsophisticated customers. In this case, a kit that shows the customer exactly what they need for system manufacturing may be required. It is up to marketing to dictate to engineering what is needed.

REFERENCE PLATFORM MRD

Many companies actually require marketing to develop an MRD to specify the application platform used for product launch. The MRD must highlight what is important to include from a board feature standpoint. In markets where a reference design or manufacturing kit is required, an MRD is absolutely necessary since the features and capabilities of the application must match the requirements of the target market segment.

Also in planning for application boards, marketing should be prepared to instruct engineering on the number of these systems needed and over what timeframe. Once again, this forecast depends on the target market and the product. Certain markets may have as little as five target customers, others may have thousands. Marketing needs to forecast their needs and let engineering know how many systems they need to build. The earlier marketing can provide this the better. In some cases application boards may require certain components that are hard to get or may be on allocation. Marketing should provide engineering with plenty of lead-time to ensure that all supporting components arrive in time to complete the boards on time for product launch.

Depending on the target market and product, it might be necessary to show customers the Bill Of Materials (BOM) estimates so they can calculate the total system cost. Usually, system BOMs are required for highly integrated SoCs. Highly integrated SoCs by definition have integrated most of the system components on silicon. In these cases, customers must evaluate total BOM to determine what solution gives them the greatest savings. If BOM is important to a product line and market segment, marketers must ensure that engineering develops a reference design and a manufacturing kit with the lowest possible BOM costs. Once again, an MRD can help engineers specify and understand the target BOM necessary to win business.

Certain products also target markets that are very sensitive to the performance of silicon in the actual system. For example, a semiconductor company might develop a tuner for satellite application that must meet certain signal to noise and other related requirements. In this case, marketing must learn what critical design parameters exist and ensure that performance data is available to help promote a product in these segments. Performance data is usually gathered once a chip is complete and running in an actual system. By the way, if performance is important to a certain market segment, then the original MRD should call out these performance parameters.

In planning a product roll out, marketing should also keep in mind that software is required to make most demo platforms, evaluation boards and reference designs work. Marketing working as or with program management should keep that point in mind as they prepare their product launch plans. Software engineering must be kept in the loop on all product application discussions.

Marketing is also in charge of ensuring that sales and Field Applications Engineers (FAEs) receive the necessary training they will need to promote and support products to customers. For large semiconductor companies, this may be an involved process. Marketing may have to ask engineering to help train FAEs since this usually involves technical issues beyond marketing's knowledge. More discussion on the marketing and sales relationship is highlighted in the following chapter.

PREPARING LICENSING AGREEMENTS

In addition, a marketing manager must pay special attention to the readiness of all legal documents that customers must sign before getting product collaterals. In the old days, semiconductor components companies created data books that gave their customer all they needed to design in products. However, current semiconductor products are more complex and in most cases contain third-party software and hardware IP. In addition, companies guard product datasheets (that could contain as many as 1,000 pages of information) with greater care than before. Marketers must clearly communicate to customers what legal documents they need to sign and return to receive a complete set of product collaterals. Typical documents include non-disclosure agreements and software license agreements.

Marketers should never let legal staff unilaterally negotiate terms of legal contracts. Lawyers are paid to deliver ironclad contracts and are less likely to be sensitive to resolving contractual negotiations quickly. This could result in product design in delays. Lawyers need marketers to help them prioritize business issues that can come up in a legal negotiation. Marketers must continuously balance legal and business issues in order to move lawyers forward during tense business discussions. A lawyer's job is to give marketing the worst-case scenarios as dictated by terms in a legal contract. Marketing must decide how much risk is appropriate given a business opportunity or situation.

Finally, marketing must ensure that a new product is set up in the product order system of the company. Typically, a product will be assigned a part number and in some cases a separate order number. Also, a product will have certain feature options or bond-out options. These options may require a product to have specific ordering instructions or unique product codes. Marketing must work closely with operations people to ensure the proper set up of this process in the order management system.

PROMOTION PLANNING

As discussed earlier, a semiconductor chip is usually sold to a business that uses it to make an end product. Activities to promote these business-to-business sales therefore will not look like the same promotional activities associated with consumer products. Consumer products will rely heavily on general advertising campaigns and the development of brand equity. Semiconductor products tend to have a much more narrow promotional target.

That said, as we noted earlier, Intel is developing a brand and is heavily involved in general advertising. For semiconductor products in general, promotional activities can vary greatly depending on the number of target customers, the target market, and the application. A new general commodity product may need more advertising in electronics magazines than a highly integrated router chip specifically built for a small number of customers. Each situation is different and marketing must determine the right mixes of promotional activities required to get the word out that a product is available to the market.

Typically, it is useful to have some type of quick overview of a product that can be passed out at customer meetings, trade shows, and other venues. This could be a product bulletin or some sort of short product brief. Marketing is usually in charge of writing these documents which are useful in situations where a product has multiple customers. If they are not too revealing of the chip IP, they can also be posted on an external website for interested customers to download. Of course, keep in mind that your competitors will also have access to these product bulletins, so it is a good idea not to include too much detail in these documents.

Marketers must be prepared to speak about their products in front of customers, journalists, financial analysts, and management. Therefore, a well thought out product presentation is mandatory. These product presentations can include a roadmap, a description of a new product including block diagrams and feature lists, as well as discussion of the product's benefits. Marketers spend a great deal of time speaking and evangelizing their products, therefore the development and maintenance of a product presentation is a large part of marketing activity.

Depending on the product, a press release or media advisory announcing the new product might be necessary. Semiconductor companies use press releases for multiple purposes. A press release is intended to inform customers about a new product offering. Companies also use press releases to signal the investment community about the prospects of a new product. The press release also serves to showcase the overall brand image of a semiconductor company in the eyes of the industry. Finally, semiconductor companies also use press releases to show off to their competition.

DEALING WITH THE PRESS

Along with a press release, marketers may also consider setting up a press tour to discuss a new product's features and benefits with industry trade journalists and research analysts. Evangelizing the product to journalists and analysts can result in more press coverage. It is also advisable to speak with influential journalists and analysts who cover a particular market segment. They can provide feedback on positioning of a product and ask questions that may reveal information about the competition as well. This also allows marketers to practice their marketing pitch before presenting it to a large audience. Usually journalists and analysts are less technically savvy than semiconductor customers. Hence, these meetings are good test of a marketer's skill to see if he can convey the advantages of his product and road map.

Marketing usually works with Public Relations (PR) people when developing press releases and press tours. Working with PR experts is a good idea since there are pitfalls in dealing with the press. A poorly managed press tour can be more damaging to a company's reputation and a product launch than just simply staying at home. Good PR people will know how to prepare a press kit, arrange meetings with journalists, and help a marketer avoid missteps during the meetings.

After setting the initial product positioning using the press release, marketing should expand and articulate that positioning in interviews with industry analysts and journalists. If a message is compelling, a marketer can expect to get some good write-ups in various industry trade journals. Press tours also offer a chance to develop relationships with key industry trade journalists. Marketers should select their words carefully during interviews and avoid tangents. Tangents invariably lead to statements that can be taken out of context. Never mention the competition unless specifically asked, even then avoid discussing the competition as any discussion gives them credibility. Well ahead of press interviews, marketers must prepare to make sure they clearly articulate the features and benefits of a new product, as well the high-level product positioning.

Along with a press release and press tour, once a new product is ready for promotion, a marketer must visit his key customers to promote the availability of the new product. In some cases, customers will be eagerly waiting for product samples if they have been sold on the product earlier. In other cases, customers will not take interest in a product until they see actual silicon. Face-to-face meetings are also an ideal way to demonstrate the functionality and performance of a new product. Sometimes semiconductor sampling plans have to coincide with key trade shows. In such cases, marketers must concentrate their promotional activities around show plans, which must include comprehensive customer meetings. Planning early for these types of events will help marketers maximize their effectiveness and promotional success.

INTERNET PROMOTION

Semiconductor customers are savvy users of the Internet. Hence, new product promotions should include an on-line strategy. Most semiconductor companies have a website which promotes the company and its products and services. The Internet can be used extensively to make product collaterals available to customers. If necessary, provide secure download capability to distribute sensitive materials such as datasheets, product briefs, and licensing agreements. In some cases, marketing materials such as presentations or feature benefit analyses are also appropriate for web distribution. (Graphics controller companies such as NVIDIA incorporate a comprehensive use of the web to communicate features and benefits of new chips to not only their chip customers, but also serious gamers.)

With password access, marketers can also provide detailed product support materials as well as software to facilitate and improve technical support. Password enabled portals can also be set up by marketers to create a customer specific environment. For example, say the Acme Semiconductor Company has built an SoC that targets several larger CE vendors including Sony Corporation. By providing a password to Sony engineers, they can go into a special site that provides information specific to the Acme-Sony relationship. Sophisticated technical support tools can also be added to help record and track technical questions that Sony generates during a design in process.

Similarly, new product information should also be posted on an internal website. This would allow sales people and FAEs to have access to the latest documentation, software modules, application board release schedules, application notes, and other design support materials. Using Web-based support tools could eliminate the need for creating hardcopy datasheet and other support documents.

ESTABLISHING PRODUCT PRICING

One of the most important decisions associated with product launch and promotion is where to set the price of a new product. Usually, prices provided in a press release are called "budgetary". This convenient description allows semiconductor companies to set an initial price point that provides a rough estimate of price to the market. In most cases, the budgetary price is set high to allow for margin protection as price negotiations and market forces drive the price lower over time. Most semiconductor customers know that budgetary prices are simply "pre-negotiated" prices and will not get offended if they seem a little inflated. But there is no implied promise that actual prices will be lower. Market dynamics can also push prices up after a product is introduced.

In a garage sale or on eBay, regardless of how much someone wants for a specific item they are selling, the value of that item is largely decided by how much buyers are willing to pay for it. The law of supply and demand regulates market price ranges, and the exact settlement is a result of negotiation skills between buyer and seller, as well as other potential factors.

Setting prices for semiconductor products follows the same principles. There usually is a general market price for each category of products. If a product is in short supply, then the price may be very high, if it's a commodity and readily available, the price is probably very low. Marketing is responsible for extracting as much value from the marketplace as possible for their products. The value of a product depends on customers' perceived value of a product's features and benefits. The economical value that a chip brings to a system design shapes the customer perceptions of a product. In other words, a customer's pricing decisions primarily depend on how much cost savings a chip brings to end products.

A marketer should never be ashamed to demand a high price for a product even though the cost may be low. If the product brings value to the market, the innovation and investment into the product should be rewarded. For example, say a marketer introduces a new product with a manufacturing cost of $5 to a market segment and he charges $1,000 for that chip netting $995 profit. Should the marketer feel ashamed of this price? The answer is that it totally depends. It is possible that the $1,000 chip saves customers $10,000 in system BOM costs over older technologies, in which case the marketer might be undercharging for his chip. The product has brought true innovation and value to the market, and profit is the reward for this innovation. Everyone is made better off.

In another example, say a chip for medical equipment costs $10 to make. Assume also that this chip replaces a medical device component that costs a customer $5,000. Assume as well that there are no competitors. What price should be set? The answer: whatever the market can bear. In this example, the price cannot be higher than $5,000. Where it lands below $5,000 may depend on customer switching costs, negotiation skills of the buyer and seller as well as many other factors. The point is, if investment, know-how, and ingenuity resulted in a chip that potentially can command a great deal of value, a marketer should never be afraid to request compensation for that value in the price.

Marketing should never use chip costs as a starting point to determine chip price. This is even true when chip prices are at or below chip costs. The case of setting price below costs is a debatable strategic decision. The price is always determined by market factors and negotiation skills, the decision to continue in a business is a different question. Either way, a marketer should never start with product cost when setting the price.

Finding out the cost and adding a "fair margin" to set the price is ludicrous, a marketer may very well be "leaving money on table".

In most cases, customers have alternative products they can choose from. It's rare that chip customers depend on a product from a single supplier for a long period. In addition, semiconductor customers are usually savvy and know roughly how much it costs to manufacture a chip. In these cases, the price of competing products will set the price of a product, along with mutual negotiation skills and the specific feature benefits of a chip.

It is possible for a marketer to demand higher prices than the competition based on a perception of higher quality or by providing a complete hardware, software, and system solution. Customers also pay more if they know they can get their products on time. By utilizing a good supply-management system, a marketer can help his customers move more toward a just-in-time inventory method and lower their product inventory costs. Since it brings value to the customer, it is conceivable that this value can be compensated in a higher chip price.

illustration 20: product pricing guidelines

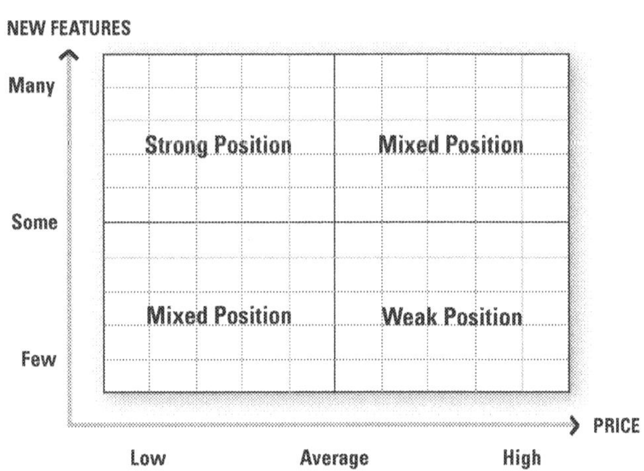

In general, marketers should review and understand the dynamics of their target market before setting chip prices. It is never a good idea to decide blindly on a price. Also, it's important for marketing to convey pricing guidelines to sales before a product is launched. Because, the first question customers will ask after being introduced to a product is the price. Sales people will feel more confident quoting price to customers if they understand how a price was arrived by marketing. Price justification will also help keep salespeople from quickly lowering the price when a customer starts to negotiate. Price is a necessary factor to win deals, but it is not the only element that determines whether a design will win in the marketplace. Marketing must clearly articulate and demonstrate the features and benefits of their products, and how those features provide real value for customers.

On this note, it is important to mention that traditionally semiconductor marketers have not been good at product promotion. Many marketers feel more comfortable relying on their engineering backgrounds and focus on promoting product features. They spend less time explaining how product features benefit an application.

SUMMARY

For some reason, semiconductor marketers do not spend much time planning product launches. This does not make sense given that a product launch should be an exciting time for a marketer. After all it is a time when years of hard work finally result in something a marketer can market and sell.

Several months before a product is launched, a marketer should produce a product launch checklist, which outlines what specific product support, promotional actions, and deliverables are needed. If certain supporting material is required like a reference design or special software for demonstrations, then these should be specified early in the development process. It is advisable to write an MRD to outline what is required for product support before a product launch.

Promotional planning should include all the necessary product presentations and promotional documentation. If a marketer decides to do a press tour before a product launch, then he should be ready with a real and convincing story surrounding the product launch. PR people should be on hand to help with dealing with the press.

Marketers should carefully set their initial market prices in a way to maximize profit. Marketers should never base pricing on product costs, rather on product value. The timing of a product launch is also crucial to the success of a product. Most chip companies tie their product introduction to an industry event, such as a trade show or announcement of a new end product. Marketing must articulate early in the development process on the product launch schedule.

Marketers need to spend as much time planning the launch as possible. Similar to a military operation, careful planning will help ensure a successful product launch.

After a product is launched, attention must turn to making money. It's hard to imagine that after all the work required just bringing a product to market, a lot of work remains in the form of making money. This can test the mettle of most marketers especially those new to the field. It is one thing to be able to have long-term strategic ideas, but it is certainly impressive when those ideas turn to buckets brimming with cash.

As we know, it's the job of marketing to define products for the right markets, manage product development processes, develop product launch and promotional plans, set product pricing, and determine the best channels for selling and supporting the products. However, to convince customers that they should buy a product from among the alternatives, marketers need to work closely with sales. The sales organization's job is to analyze the dynamics of target customers' buying processes and manage the selling process. This chapter examines ways of working with sales groups effectively.

SALES FOCUS

Sales groups are responsible for delivering a top-level revenue plan to corporate management. Sales groups will also work with marketing to develop revenue plans for specific business units, though sales will not likely tie their targets to specific products preferring to set revenue goals based on a product mix. In addition, sales organizations, depending on how they are rewarded, will normally pursue the easiest sales opportunities. These dynamics often make it difficult for individual product line marketers to get a fair share of sales' focus especially on a new product line.

One of the most important challenges for marketing is to convince salespeople that a product is worth their attention. In effect, a marketer must sell his product line to sales. Marketing must not only sell the benefits of the product and the promises of the market, but also the competitive advantages that will help guarantee sales.

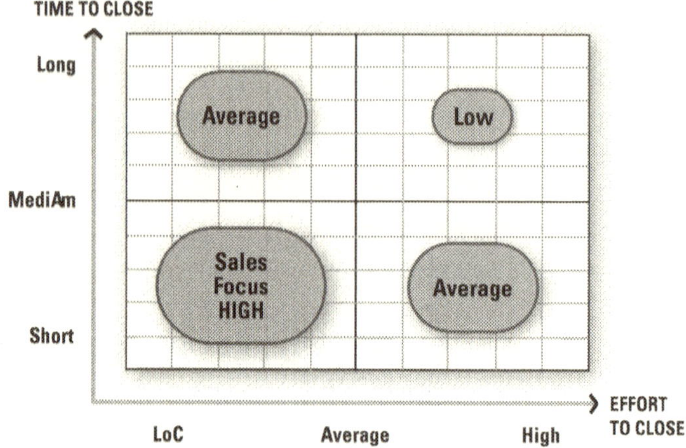

Having a good product-support strategy also helps to sell a product line to sales. Salespeople like to have their job made easier for them. They appreciate the availability of product documentation, reference designs, and support staff to help them in the design-win process. Salespeople, then compare and contrast the product line with other product lines they have to support. Needless to say, they will gravitate toward those product lines that show the most promise for sales.

Good marketers treat salespeople as partners in developing the understanding of the marketing landscape — salespeople rightly feel they have a pulse on the market and will appreciate being asked to participate and comment on product strategy formulation. Marketers should use every opportunity to determine how salespeople think about a market and the product strategy aimed at that market. In addition, marketing should take the time to explain their roadmap and product strategy to sales early in the product development process. Salespeople are on the front line of attack when it comes to a product line. It is critical that they agree on strategy and the benefits of the product line. Salespeople can also let marketing know when features of a product or an element of its support is inadequate or missing.

It's also crucial that marketers provide sales with competitive data as well as comparisons of features and benefits of the products against the competition. A sales person must know the competitions' strengths and weaknesses. Marketing must never provide sales with misleading or inflated information. Misleading information can lead to sales losing credibility in front of customers. Along the same vein, salespeople need to know the truth about the status of products and schedules, so that expectations are not built up with customers that turn out to be false.

SUPPORTING SALES

To be successful at their jobs, salespeople need all the support they can get from marketing to help in the initial stages of customer engagement, as well as through various stages of a design win. This means salespeople need direction with regard to an entry strategy into an account, product presentations, and marketing collaterals to help showcase products. As mentioned above, a product launch must include giving salespeople extensive training on key selling points of a product.

Once product training is complete, marketing and sales should travel together to visit key customers. This will allow salespeople to see marketing make the product pitch and learn how marketing is positioning the product to customers. Salespeople will also learn first hand how marketers deal with customer questions and criticisms. By giving effective and persuasive presentations to customers, marketing can gain the respect of the sales staff. Salespeople must see that marketing has done its homework and that more importantly the product excites customers' interest. After each customer meeting, marketing should ask salespeople for feedback. Sales can provide an honest opinion about the product pitch and the customer reaction.

In the early stages of a product launch, while interacting with salespeople, marketing must avoid feeling a little arrogant given that they hold most of the cards in terms of product and market knowledge. This arrogance is detrimental in several ways. At a minimum it sets up a bad expectation, one where sales convinces itself that only marketing has the power to sell a new product. Finally, marketing's narcissism will detract from the process of transferring knowledge to sales. In fact, marketing should not only sell its product line to sales but also give them as much knowledge as possible so that they can serve as independent disciples for a product line.

For these reasons marketing must treat the sales staff with respect. Marketing must make sure to return phone calls and emails related to sales activities promptly. In addition, marketing must deliver on action items and help salespeople succeed. Those marketers who feel that they know how to sell and could do a much better job than the salespeople, should think about moving to the sales organization. At the same time, marketers should not be passive when they see that salespeople are not doing enough to promote a product.

In fact, marketing should watch salespeople closely. Sales should be fighting for revenue. They should not be allowed to give up too early in winning customers. Salespeople who constantly demand lower prices or more features to sell a product, where competitors are happily selling theirs, may need additional help in product positioning.

Finally, marketing must never show weakness in front of sales. If a marketer looks unsure of himself or overly worried about the competition, it is unlikely that sales will pick up the flag and storm the castle walls. Marketing must keep a positive attitude, one of cool confident leadership. Marketers should reserve discussing personal concerns, complaints, and worries about their products with close colleagues, or their spouses.

CUSTOMER NEGOTIATIONS

In many cases, marketing and sales work together to help close deals with customers. Often marketing is brought in to help negotiate prices with customers. As we mentioned above, price is determined by supply and demand as well as the negotiation skills of the negotiating parties. Price determination is also a factor of economic value and BOM savings a customer enjoys by using a product. Finally, the costs associated with switching over to a new product will also affect its price.

If salespeople are not well equipped to sell the value of a product to customers, they will naturally gravitate towards the one factor everyone understands, price. Of all the marketing tools available (price, product, promotion, and place), price is the easiest tool to use. This is especially true with a product with a higher elasticity of demand (which means that a lower price will lead to greater sales.) Products such as memory are included in this category.

However, for the majority of semiconductor products, price should not be the only deciding factor in guaranteeing a design win. Marketing should ensure that salespeople know what a product does for customers and how much customers should be willing to pay for it. In extremely competitive markets, price may become the preeminent issue. However, even in those markets, marketing must supply sales with weaponry to fight for the best possible prices. These weapons may include BOM, competitive information, or feature benefits analyses. It may also include non-product related factors such as industry relationships, patents, or other business factors among other possibilities.

It is a good idea for both marketing and salespeople to receive as much negotiations training as possible. There are excellent courses available. The fact is that expectations and attitude play a huge role in the outcome of negotiations. Courses can teach how to use expectations and attitudes to help negotiators achieve the best possible outcomes. Good courses will also teach how to prepare for negotiations, and provide tips and tactics on maximizing negotiation success.

VOLUME-BASED PRICING

Negotiating skill is extremely important especially in the semiconductor industry. One problem with negotiating semiconductor prices is that normally negotiations focus on price per unit. This serves to hide the true impact of a price concession. For example, say a chip will sell ten million units over a specified timeframe. A ten-cent reduction in per chip price reduces the value of a deal by $1 million. $1 million is a lot of money, but ten-cents is not. Marketing should be ready to point this fact out during price negotiations.

A good first step in negotiations is to provide sales with a price book. Prices have more credibility when they are written down in the eyes of sales and customers. Depending on the product and the market, marketing might also provide a volume-based discount matrix of prices in the price book. Since most customers know that price breaks are possible based in return for volume purchases, they will ask for this information. Request for volume breakouts is the beginning of price negotiations. If prices are written down on paper, there is more chance that customers will take the prices more seriously.

Even though haggling will decide the final price of a product, it is always a good idea to develop and maintain a volume-based price list for products. How a price is broken out as a function of volume depends on the product and the market. A typical SoC product might show volumes at 10K, 100K, 500K and greater than 1M unit breakouts. The volume discounts can trigger at either cumulative volume over a specified time frame, volume per purchase order, volume based rebate, or some other formula.

In some cases, marketing may need to provide a "public" price for press releases or to deal with general requests for pricing. For a public price and for budgetary price to customers, marketing should use a lower volume price such as the 10K unit volume price off a pricing list. This way margin is left in the price before negotiating a specific deal. Also, it's a good practice for marketers to record any pricing provided to sales staff in a central knowledge repository. This allows marketers to refer to sales quotations provided to various sales regions and customers when providing pricing for new opportunities.

In most cases, marketers do not need to share product cost with sales. However, in highly competitive cases it is useful to make sales aware of cost and margins so they are more sensitive to the effects of price reductions. By knowing a product's cost, salespeople may work harder to get more money from customers. If marketing is convinced it can close at higher prices for whatever reason (for example, past customer experience, or inside knowledge about a customer), then marketing must explain the reasons for the pricing strategy to sales.

PRICE REBATE

At certain times, marketing might have to pay pricing incentives and rebates to close a design win. Sometimes these incentives can get complicated especially if they involve quid pro quos associated with the deal. Marketing should always clearly document these deals and pricing mechanisms and ensure that sales and the customer are clear on how the deal will work. Marketers should also rely on upper management whenever special deals are being proposed to ensure revenue expectations are met.

Sometimes to close a deal, either a buyer or seller will request a contract that specifies all aspects of product support, sales, and pricing. To develop a solid contract, marketers must work closely with salespeople. Salespeople can bring the right people from the customer side to negotiate a fair contract. In addition, salespeople have to support customers after contracts have been signed. Therefore, marketing must make sure sales is involved in all aspects of contract negotiation. Senior salespeople, in fact, should take primary control of contract negotiations. During contract negotiations, marketing together with sales may have to visit their customers often. Salespeople must be present at key customer meetings to provide the local level of support needed to make customers feel like they are getting focused, continuous service.

Lawyers should always be part of a contract negotiation. As we discussed above, marketing must work closely with the legal staff to ensure that contract negotiations do not get bogged down in legalese. It usually takes guidance from businesspeople to help conclude contract negotiations in an acceptable timeframe.

A book can be written simply on the topic of semiconductor pricing and contract negotiations. There are many tactics, pitfalls and strategies to look out for when negotiating with customers. These details are outside the scope of this book. That said marketing should always build expectations into sales to fight for the highest possible prices. In addition, sales and marketing should work closely together during price negotiations. A salesperson that closes deals by quickly agreeing to a low price without consulting marketing can end up costing a company large sum of margin dollars. Remember in most companies, salespeople are not responsible for products' profit margins — that's marketing's responsibility. Therefore, marketing must take an active role in setting product pricing and price negotiations.

ACTIVE SALES TRAINING

Many companies depend on an annual sales meeting event to train their salespeople. This method of training may not give salespeople the tools they need to do their jobs. Salespeople must get at least a quarterly update on product lines, including the competitive landscape surrounding it. The quarterly reviews can be done using Web-based presentation programs to minimize training costs.

Customer visits are a great opportunity to train sales. Often, a customer visit requires that a marketer spend a lot of time with a particular salesperson. The marketer should make use of this time to provide information and insight to the salesperson about the product line. It is also a good time for the marketer to hear from the salesperson about how customers are receiving his products and hearing about the competitions' activities at various accounts.

Marketers must keep in mind that salespeople may have several products to sell and will need to be updated frequently on product schedules and general competitive landscape. Salespeople should always be prepared to argue the competitive advantage of a product. Using an internal website as mentioned above is a great way to post the latest competitive information, news, and success of a product line. Email newsletters are another way to do this.

If a company publishes internal marketing or sales collateral material, marketers must ensure that their product line is covered properly. Salespeople are always comparing marketing activities among product lines. A marketer must provide every reason to salespeople to give the attention his product line deserves.

As part of a sales training program, marketers must develop a team spirit with sales. Team spirit might take some time to establish. Marketers need to work closely with top-level sales managers to arrange team-building activities such as a golf invitational or other sporting events. Also, marketers must participate in sales conventions and outings, when invited. For example, if there is a sales golf tournament before an annual sales meeting, marketing should seek to be part of the event.

If marketing cannot arrange for official team-building activities, good marketers should look for ways to interact with salespeople after official working hours. Industry shows and conventions, post-customer meetings and sales visits to headquarters, offer the chance to take a salesperson out to dinner, or other socializing. Socializing with sales is a good way to develop closer relationships that help establish team spirit.

ROLE OF DISTRIBUTORS AND REPRESENTATIVES

Many semiconductor companies use distributors, especially companies doing sales overseas. In fact, in some emerging markets, the partnership of a reputable distributor is mandatory. Distributors can extend the necessary credit to customers and help guarantee that invoices are paid on time. Salespeople should be responsible for managing distributors and representatives (reps) in their respective areas.

If distributors and representatives are the only link to a particular market, marketing must take extra care to support them as they would be the internal sales groups. This is an especially important point for higher-end SoC product sales that require much training. Many semiconductor marketers would like to avoid dealing with distributors and representatives. After all, marketing may judge that support of reps and distributors are sales' responsibility. But this is not the right attitude.

If a group is assigned to promote a product they deserve the attention and support necessary to do their jobs. Marketing may question the need for distribution and reps to promote a product. In fact, many larger companies may employ both inside sales and reps to promote products in the same region. In cases like this marketing can voice concern over whether reps are needed. However, once a decision has been made, marketing must work closely with these groups to ensure they have the tools to succeed.

In working with distributors and reps look out for the creation of gray markets. To do this, match the volume demand of each customer with the total demand given to you by distributors. If there is a mismatch, products may be selling through unauthorized channels at prices lower than established regional pricing, creating a gray market for products.

ROLE OF FIELD APPLICATION ENGINEERS

In most semiconductor sales organizations, Field Application Engineers (FAEs) report to regional sales managers. The role of FAEs is to communicate the technical needs of customers to inside application engineering groups, support the technical needs of customers during design-win activities, and provide support during the production phase of customers' products.

When marketing creates its annual sales training calendar, it should also include specialized training for FAEs. The training should focus more on the inter-working of products and perhaps less on the competitive positioning. FAEs will need detailed hardware level and software training necessary to support customers.

Training FAEs is a great way to get engineering level feedback on a product or product line. FAEs are a great source of information on how customers perceive a product versus those of the competition. The FAEs are the primary contacts with the customers' engineers. Unlike business-level contacts, engineering contacts tend to share more information, making FAEs a valuable sounding board — not to mention a great source of intelligence on products and competition.

Marketing must ensure that internal application engineering groups give first-rate service to FAEs. Well-trained FAEs can make or break the success of a design in. FAEs often are a salesperson's right hand when

CHAPTER 8 >

it comes to winning and keeping deals. Taking the time to train FAEs is critical. Consider having FAEs come to headquarters or the appropriate research and development center to work closely with your internal application engineers for extended periods. This is a great way for FAEs to learn firsthand how to deal with the complexities of a product design in an end application.

SUMMARY

Many semiconductor companies rely too heavily on marketing in the sales process. We strongly believe in the need to empower sales to do what they do best — selling. In fact, this chapter did not address the sales process per se; after all, this is a book on marketing, not a book on selling.

Marketing must avoid making salespeople into highly paid liaisons who do nothing more than set up meetings and chauffeur marketers to customer locations. When this happens, the selling process falls on marketing, and this means key marketing functions, as well as sales potential, will suffer. In our view, this situation is mostly marketing's fault. In fact, it is not uncommon for marketing people, who are naturally steeped in product knowledge, to try to take the glory of the product sale. Certain marketing dilettantes want to show the world how invaluable and important they are by hoarding product information and sharing little — forcing the sales team to depend on them to make the sale.

Marketing must work to maximize the sales team's effectiveness by empowering them with information and knowledge. In return, marketing must expect salespeople to work hard to close business. Nevertheless, if there are salespeople who are unable to articulate a marketing strategy and deliver revenue, then they need to be dealt with. True market success comes when the whole team is fighting for revenue. A good product that has been positioned well for a market segment and is supported by an aggressive sales team will start selling and turning a marketer's ideas to money.

CHAPTER 9
providing technical support ›

Developing the right products for any segment of the semiconductor industry is complicated and expensive. Companies routinely spend millions of dollars bringing new product ideas to market. Many of these same companies, after spending the money to create the products, often are reluctant to expend additional resources on providing adequate support for these products. Companies routinely underestimate the effort and expense required to win and keep customers.

As the complexity of semiconductor products grows, so does the need for a higher level of technical support. In addition, as the industry moves to smaller fabrication geometries, more companies will introduce SoCs, making it almost impossible for customers to design in these products without a lot of support. Technical support is a function of the product complexity, its position in the value chain, and the product distribution channel. Support strategy also includes considering how to organize an application engineering group, and the use of third-party hardware and software developers.

As we discussed, setting up a support strategy must be part of product launch planning. This chapter focuses on technical support activities in more detail during a product launch as well as the long-term technical support of a product. Ensuring that technical support is set up well in the initial phase of a launch will help the success of a product through its lifetime.

THE SoC ERA

In the old days, semiconductor companies produced a data book, which contained datasheets for various component products and a few application notes that showed systems designers how to use the products. Today, a single SoC datasheet can easily be 1,000 pages or more. In addition to a complex datasheet, customers must also learn how to utilize, or sometimes write, an entire software application necessary to create an end system solution. Therefore, it is important that marketing define how much support in terms of documentation, evaluation boards, application software, and resources is needed before products are released to customers.

It is possible that many companies can duplicate a product and compete on price. However, in the end, the one thing they can't easily duplicate is an excellent technical support team. We all have good and bad customer experiences. Some researchers believe that when a person has a bad customer experience, he will tell more than 15 people about it. However, when a person gets great customer service he tells no more than three to five people about it. This implies that bad customer support is potentially fatal to success of any company.

The high technology industry is dynamic, and companies cannot afford alienating customers through poor or non-existent support. Sometimes companies that are ahead of the pack in the technology race can easily lose sight of this point. The complex nature of semiconductor devices gives the designer of products, and those who market and sell them, a false sense that their products are so far ahead of the pack that customers will buy them, no matter what. In some cases, these companies convince themselves that no one would or could ever attack their product offering. This is a shortsighted view, because the industry is full of innovative, smart people who find ways to create similar or better products. Therefore, marketers must always treat customers well. All members of an organization should also realize their job is to make customers successful.

SYSTEM VERSUS COMPONENT SOLUTION SUPPORT

Most semiconductor companies have both component- and system-level solutions in their product portfolio. In the past, components business usually did not require the sale of software along with chips. However, this is slowly changing. While it is true that very simple components still do not require software, many now require some level of software driver support. In general, components are easier to support than complex SoCs. Nevertheless, they still usually require good application notes and technical people on hand to answer questions and help with application design.

In some cases, a complete board layout solution is required to get a component to work correctly. For example, RF components require very tightly specified layout requirements. Minor layout discrepancies for an RF component could prevent the final product from performing. Technical support must ensure that customers consistently achieve circuit implementations that are within necessary tolerances.

On the SoC side, technical support issues usually revolve around software. In an ideal world, it is always desirable to have a semiconductor SoC solution that customers end up writing the software for, striving for their own differentiation in the process. This ideal world is becoming increasingly rare. Semiconductor companies developing system-level products typically need to provide system-level software to round out their product offering.

illustration 22: application support model

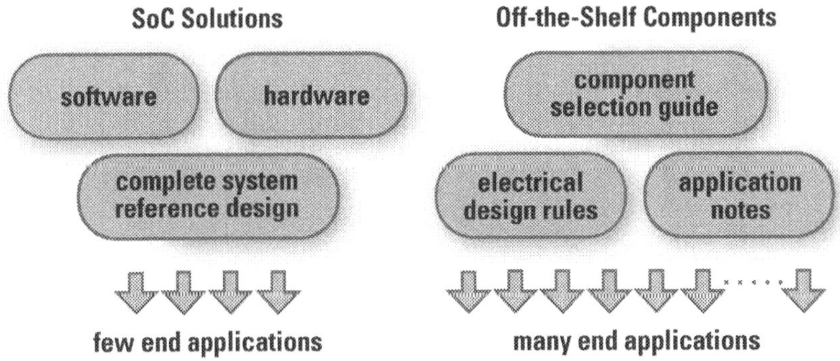

Some semiconductor companies end up writing entire system-level software applications for their products. This is akin to Intel developing a complete operating system as well as applications software to work on their Pentiums. In most cases, this is not an optimal strategy, but it may be necessary to do, especially for new entrants attacking entrenched incumbents. However, it is not a great position because software support never ends. By taking on offering a complete system solution, a semiconductor company signs up for complete support of the entire system before and after its release to end customers.

What makes this situation even worse is that semiconductor companies have not been very successful in recouping investments made in software. Typically, customers who use semiconductor products are not used to paying for software provided by semiconductor companies. In addition, semiconductor companies, which have been hardware-oriented, have therefore, been slow to recognize how to market and sell their software solutions.

Supporting an SoC requires having key application engineers who understand software development and support. Product application groups should develop full-featured bug tracking and software release tools, just like real software companies. Application groups should also divide the support tasks between hardware and software to give the support engineers the time they need to become experts in each of their respective disciplines.

It's crucial to communicate clearly to customers how much software and support come with products. Customers have a tendency to assume that certain tasks are the responsibility of the semiconductor company. To avoid misunderstandings, semiconductor companies must make use of written memos that clarify exactly what they plan to deliver in terms of software. Of course, it goes without saying that credibility depends on delivering.

In addition to providing software and hardware support to OEMs and ODMs during the development of systems solutions, field support after products are released to end customers might be necessary. For example, an SoC chip vendor who is providing a graphics controller board to a PC OEM, via an ODM, may have to provide a direct customer support line to help with driver related issues.

ROLE OF FIELD APPLICATION ENGINEERS

Most semiconductor companies' application engineering groups, which have the charter of supporting customers, are part of marketing groups. Applications groups are entities that, for most part, set the tone for how an entire organization treats customers. If a patient service-oriented person runs an applications group, then customers will view the whole company as a service-oriented company. On the other hand, if the person running the applications group is impatient, cannot handle multitasking, and does not feel responsible representing the company, the result is unhappy customers.

It may be advantageous to locate support people remotely to be near customer sites. By doing so, a marketing manager can better monitor how customer issues are recorded and resolved. In addition, locating support people near important customer sites will allow for localized and timely customer support. Efforts should be made to promote close cooperation between support staff in a remote location and design centers. This would be the key to improving the flow of information to and from customers.

However, in many companies, application engineers in the field report to the sales group and may have to support multiple product lines. If the application engineers in the field are not dedicated to one product line, their ability to build the product expertise necessary for proper customer care will be limited especially for large SoCs.

A marketer may be able to solve this problem by having field application engineers work at a product development site for a few months to gain expertise and interest in a product line. Field application engineers will often favor a product line they are familiar with technically. In addition to training, the marketer can also ensure that field application engineers are supported by a trained and responsive set of factory application engineers who are dedicated to a product line.

Having a decentralized and highly skillful support group in the field can be crucial to the success of a product line. As we noted, this is especially true of large SoCs. It may, however, also be a requirement for off-the-shelf products. Off-the-shelf product support could be greatly facilitated by providing extensive web-based support.

If a semiconductor company has products that target various industry segments, such as PC OEMS, cable companies, or large telecommunications companies, it is probably a good idea to dedicate separate application engineering groups covering each industry segment.

SUPPORTING OEMs AND ODMs

To better align a support organization for the support of customers, it is good to understand the role of Original Equipment Manufacturers (OEMs), Original Design Manufacturers (ODMs), and semiconductor distributors in the semiconductor value-chain. As noted previously, OEMs add value to semiconductor products by developing system solutions, integrating ICs from different vendors, writing system-level software, marketing and selling the end products, and supporting customers. OEMs often assign a large group of engineers to bring new solutions to market. In some sense, OEMs do not like to see semiconductor companies develop the entire system-solution hardware and software and then give it away free, as they consider this a competitive threat.

To win business at OEMs, IC companies must have advanced and innovative products that are superior to those of competitors in features and performance. OEMs typically require complete hardware specifications as well as a generic set of software drivers and documentation. The OEM usually completes the software application stack necessary to bring products to market. To support large and important OEMs properly, marketers should consider assigning a dedicated team to support them. In some cases, a marketer may even assign one or more engineers to become part of the OEMs' development group to provide local and dedicated support. The additional step of assigning program managers to OEM product development effort might also be necessary depending on the criticality of the opportunity.

Unlike OEMs, ODMs, which are mostly located in the APAC region, often enter a market when the form, fit, and function of end system solutions are well defined. The ODMs cannot afford to define, build, and market products for the early phases of a new market segment. For example, early consumer routers were developed and marketed by established OEMs such as 3Com located in the United States. As the software and hardware for these products matured, companies like Linksys were able to enter the market by utilizing the low-cost design and manufacturing capabilities of ODMs in Taiwan, effectively lowering consumer router prices. With Linksys entering the market, the price for consumer routers subsequently dropped from $500 to about $50 in less than three years. Another example of a successful ODM play is 802.11b/g wireless access point product sector. Semiconductor vendors providing 802.11 solutions have all used the ODM phenomenon in Taiwan to provide products to end markets at a rapid rate.

ODMs located in Taiwan and increasingly in China will require a complete hardware and software reference design to minimize their development cost and time to market. In the case of China, it may even be necessary to provide them with a complete manufacturing kit — all the hardware, software and test information necessary to go directly into manufacturing.

SUPPORTING ODMS

The APAC ODM channel requires having a skillful group of application engineers in the region that can help get ODMs through their designs quickly. In fact, when it comes to the ODM model, the quality of technical support greatly influences the success of a product line. Most products that find their way into the Taiwan ODM community are mature and therefore cannot be differentiated with features. While pricing is a powerful tool to get new design wins in Taiwan and China, the key differentiator is the quality of technical support staff. Even with great prices, lack of good local support will greatly limit the success of any product in the ODM channel.

In addition to providing the ODMs with local applications support, during the design phase, it might also be necessary to give ODMs 'end customer' support . This is especially true if the products are sold to large PC

OEMs or service providers. For example, if a Taiwanese ODM is shipping a graphics controller product to Dell, then most of the field support and system-level issues may have to be taken care of by the graphics controller chip company. After all, Dell sees the chip vendor as the main IP vendor. Contrast the ODM model to the case where Dell buys its product from an OEM based in the United States, where most of the product support would come from that OEM. For example, if Dell were to buy a communication product from 3Com, most likely 3Com would end up providing the field support and the chip company behind that product.

The support costs and tasks involved in getting a semiconductor designed into an end product remain the same regardless of which group of manufacturers is behind the product design. In the case of OEMs, the majority of the software development and support costs stay with the OEMs, and hence OEMs end up charging more for their products. In the case of ODM designs, the brunt of support costs will get transferred to semiconductor companies. This means that a semiconductor product sold through an ODM channel must sell in high volumes to make it a viable business. Otherwise, the support costs associated with running products through the ODM channel make it economically impossible for semiconductor companies to realize any profit.

SUPPORTING DISTRIBUTORS

Regardless if an ODM or OEM model is used, one question that often comes up is whether or not to use distributors to sell products. Distribution is often the only way to sell products into a particular channel. For example some large OEMs insist on buying all their products from one major distributor. Or in the case of an emerging market like China, the use of a distributor helps guarantee payment on product shipments as they are in charge of establishing credit to Chinese accounts.

Distribution also has some drawbacks. While it is true that many distributors can provide local applications support to customers, they are unlikely to be able to handle complicated designs and SoCs. Most semiconductor distributors cannot afford to provide the level of engineering needed to support a SoC solution.

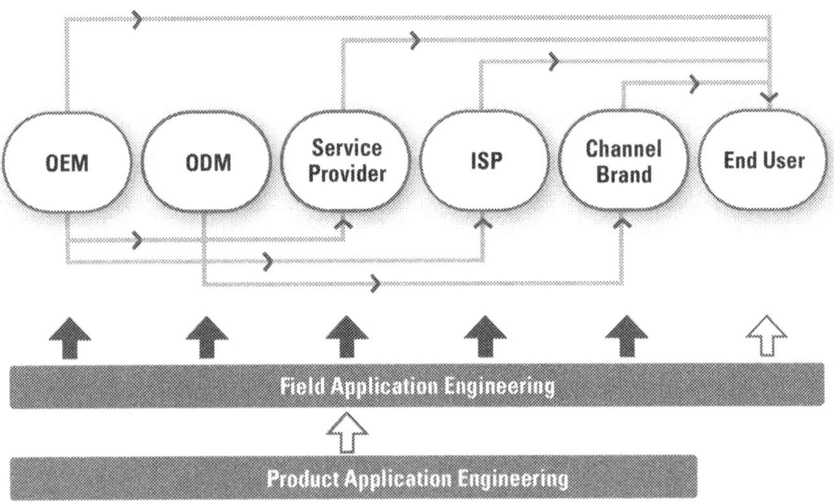

Distributors are great at maintaining customer relationships with smaller companies. This is important when a semiconductor company does not have a large sales staff that can cover all regions of the world. Distributors can find local third-party support people who can help customers with their design-related issues. For example, it might be possible to find local low-cost engineering support in China or India to bring system-level solutions to market quickly. For very complicated solutions, even this model can have difficulties since it is costly and time consuming to bring up third party support people to the level necessary to make them effective.

SUPPORTING INDEPENDENT SOFTWARE VENDORS

Depending on the complexity of an SoC, it may be impossible to internally staff enough software engineers necessary to develop software for all the potential applications of the SoC. In these cases, marketing should consider a strategy that includes the help from Independent Software Vendors (ISVs). ISVs make their money by adding value to products. A marketer, who can create a business model around the support of ISVs, has effectively created an industry around his products. This means other ISVs' livelihoods depend on the success of the SoC.

In the graphics controller chip market, the support of game developers for a specific platform is a key to its success. In fact, most of the advanced features of a sophisticated graphics controller chip go unnoticed. It is all about the success of game developers, the ISVs, that counts. Therefore, companies such as NVIDIA spend large sums of money on software tools to help ISVs develop games for the NVIDIA platform.

When developing an ISV strategy, a marketer must evaluate what tools are necessary to support the ISVs. There might be market opportunities available with the right ISV tool strategy. In addition, application engineers should be equipped to support ISVs efficiently and effectively. ISVs are like customers; even though they do not pay, they still add value to products, and that means a marketer must be prepared to give them first-rate support.

SUPPORTING RETAIL BRANDS

Consumers' need for digitizing various content (video, music, pictures, text), home networking of computers, and broadband access have driven the demand for consumer communication systems. Consumer communication products are primarily sold in the retail channel by companies such as Linksys, Netgear, D-link, Belkin, Zoom, SMC, and Best Data, to name a few. These companies mostly depend on ODMs in Taiwan and other parts of Asia to develop and manufacture their products. These channel brands have also started targeting service providers such as telecommunications and cable companies to grow their revenue. Unlike retail customers, service providers demand a much higher level of customer support.

Semiconductor companies that plan to support service providers with retail channel brands must realize that the brunt of the service providers' support cost and effort might fall on them, the semiconductor providers. The retail channel brands look for support in selling and marketing their products to the service provider to combat the marketing and sales arms of larger

OEMs. This could further pull the marketing and support effort of a semi-conductor company into supporting all technical aspects of field issue.

When introducing products, it's often difficult to switch from an OEM to ODM channel. A semiconductor company may have little choice but to actively move the distribution channel away from OEMs to ODMs, due to the changing market dynamics such selling products directly to service providers. In this case, establishing an effective field support staff to support service providers, on behalf of the ODMs and retail brand, could provide a marketer with a sustainable competitive advantage.

SUPPORTING END CUSTOMERS

Most semiconductor companies do not bring products to end markets using their own brand. Even Intel, with all its power, depends on Dell and other PC OEMs to mass-market its processors. However, in some cases, and depending on the market dynamics, semiconductor companies may decide to own the entire value chain and bring products to end markets based on their own brand name. The ramifications of having such a strat-egy could be detrimental to the long-term viability of a semiconductor company if not considered and executed carefully.

In some cases, system companies develop their own chips for their internal product consumption, and depending on the success of a prod-uct, they could even sell the product to other companies. For example, Qualcomm has been successful in developing wireless chipsets for in-troduction of its own CDMA phones, and then transitioning to become a major semiconductor company. After initially selling its own brand of mo-bile phones, Qualcomm exited that market and sold its handset business to another company to focus more on selling its chips to other handset manufacturers. There are also semiconductor companies such as ATI that develop advanced graphics controller chips for PCs that are primarily used in ATI's brand of graphics controller cards.

SUMMARY

Poor support can lead to frustrated customers and delayed product ramp schedules. A market leader that controls major share of a product market might be able to get away with poor support temporarily. However, a reputation for poor support will eventually catch up with anyone. Marketers must take active roles to ensure that their products are supported effectively. Product support influenced by a particular solution and the product distribution models.

regional market development >

In many semiconductor companies, marketers use geographical boundaries as a way of focusing on target markets with distinct market characteristics. The primary market regions are the Americas, Europe, Asia Pacific, and Japan. These regions consume the bulk of annual semiconductor production.

Depending on the complexity of a semiconductor product and the number of customers, it may be impossible for a single marketer to effectively cover all of the worldwide regions. Each of the previously mentioned regions contributes differently to the success and annual revenue of a product line. The Americas region, especially the United States, is one of the primary markets for finished computer and electronic equipment manufactured in the Asia Pacific and Japan. Taiwan, Korea, and China are the primary manufacturers of electronic equipment for the world markets. Taiwan is the leader in producing personal computer-related systems and peripherals.

illustration 24: regional market segmentation

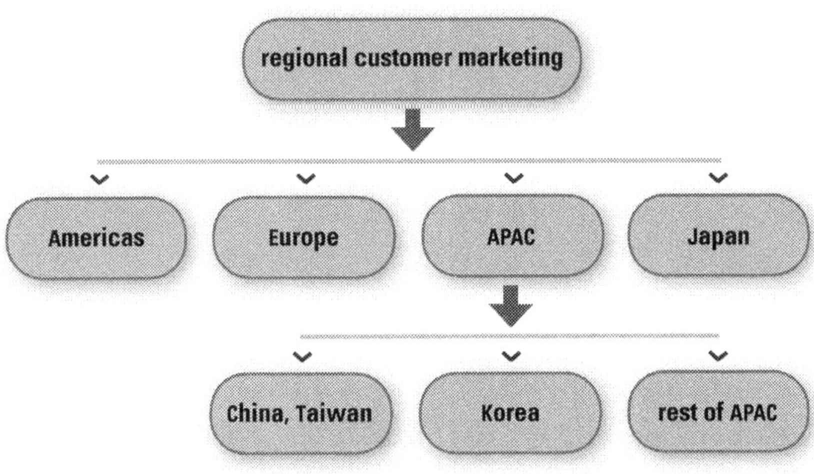

American OEMs must develop products that can bring enough margin dollars to justify their higher costs of developing and selling products worldwide. These OEMs achieve profitability by delivering very advanced and innovative products to their target markets. For this reason, U.S. OEMs like to work with and depend on semiconductor companies that are also developing cutting-edge products. The key to winning U.S. OEM business is to be the technology leader with a great vision and roadmap for the future.

The good news is that once a US-based OEM decides to design-in a chip, they have a tendency to stay with their primary chip supplier over the long term. This tendency to develop long-term partnerships with suppliers creates a market barrier to competition.

It is no surprise that U.S. OEMs pay special attention to product roadmaps. Since they prefer to work with a supplier long term, they scrutinize roadmaps to ensure potential partners are thinking about and planning for innovative solutions. OEMs want to avoid costly switchovers as much as possible. A strong roadmap will help convince U.S. OEMs that a vendor is a good bet as a source of future innovation.

As part of promoting the innovative capabilities of a company, top-level engineers and scientists must participate in various industry-standards committees, and actively produce relevant and valuable patents for a company. However, nothing speaks more to the ability to innovate than bringing new products to market quickly. There are companies in the industry known for their ability to execute and deliver products to market in a timely manner. These are the ones that tend to enjoy business with U.S. OEMs.

There are also companies known to be followers. The marketing groups in the "follower companies" often have a difficult time convincing the U.S. OEMs to design in their products. Followers are further disadvantaged because they must prove their capabilities by developing complete system solutions for customers. This is not usually the case for the market leader. OEMs are eager to get their hands on innovative products from market leaders and develop systems around them—in effect, sharing the risk in developing system solutions. A follower, on the other hand, is forced to take on the system risk to even be considered for a design. That said, a follower strategy might work in other markets, as is discussed later in this chapter, but not when targeting OEMs.

In addition to promoting products to OEMs in the U.S., a marketer might need to promote the features and benefits of a chip to the OEMs' customers. Typically, these could be service providers or large PC OEMs. By "service providers," we refer to telephone companies, cable companies, Internet providers, and direct satellite service providers. When a product targets Internet appliances, for example, a marketer should pay special attention to Internet service providers, such as AOL and MSN, to understand what products they might need. Service providers can also provide input on a roadmap and discuss what they need to lower their capital expenditures, lower their operations cost, and increase their revenue. Having close relationships with service providers also helps influence OEMs in their semiconductor selection.

A marketer might also need to promote his products directly to US-based channel brands, such as Linksys (a division of Cisco), D-link, and Netgear, to name a few. The channel brands primarily sell to retailers in the U.S. and around the world, but do very little design and manufacturing in the U.S. In the past few years, new players such as Linksys have been able to dominate the retail market for home networking products. Most channel brands have a strong partnership with several ODMs in Asia. The channel brands also influence the selection of semiconductor products by ODMs in Asia.

Besides managing the channel brands in the U.S., a marketer might also need to promote products directly to consumer electronics retailers. In some cases, a chip marketer may even decide to contribute to retail promotions to help drive demand for system solutions that use his semiconductor ICs.

Marketers need to continuously monitor the retail channel in the United States to make sure new competitors are not making inroads on their established customer base. Studying the retail channel will also provide insight into buying trends and technologies that are popular with consumers; this insight is useful in the development of a product road map.

EUROPEAN REGION

Like the U.S. market, most of the semiconductor demand in Europe is generated by OEMs. To gain market acceptance in Europe, companies might consider establishing a focused marketing presence in Europe. English is the language of business in Europe (this trend has even found acceptance in France). That said, it never hurts to have bilingual marketers to help with delicate negotiations and communications at key OEMs. If having a dedicated marketer in Europe is not possible or appropriate, marketers will need to work closely with the local sales teams to develop relationships with key European OEMs.

In the case of the PC market, recognize that American companies through their subsidiaries supply most of the PCs sold in Europe. Relationships with large PC manufacturers in the U.S. can provide a marketer with the leverage needed to drive demand from Europe for his semiconductor products.

To better promote products to OEMs in Europe, marketers should participate in various industry shows. European companies often look for leading-edge technologies that could be customized for their region. These shows can help marketing understand the product features needed in Europe.

Asia Pacific and Japanese OEMs have also made large inroads into Europe. When covering Europe, marketing must determine which of the Asian companies will succeed in obtaining significant European market share. By having good knowledge of European requirements, and ideally having the support of European service providers, marketers will have an easier time getting design wins with Asian OEMs and ODMs targeting Europe.

European service providers determine the direction of semiconductor products used in equipment they deploy. Marketers need to determine which European OEMs have key relationships with service providers. Remember that just having a design win with an OEM does not guarantee revenue for a product line. End-customer demand drives revenue. Picking the right OEMs and channel brands that know the European market needs is of utmost importance.

THE JAPANESE MARKET

Largely, Japanese consumer electronic giants, such as Sony and Toshiba, still dominate the worldwide market for semiconductor parts used in various consumer electronics products such as DVDs, digital cameras, digital televisions, and game machines. Japanese consumer electronic companies are well positioned to benefit from the digitization of content and growth in the adoption of broadband access around the world. In addition, car companies continue to utilize more electronics. Given the position of Japanese car companies in the world, it's crucial that semiconductor marketers stay close to the Japanese market if they are targeting (or plan to target) the automotive industry.

Japanese are also large consumers of electronic products such as digital cameras, mobile phones, HDTVs, and PCs. This means that a marketing team must pay special attention to Japanese consumer market trends as those trends could eventually find their way to the United States and Europe. Japanese market trends in mobile phones and home theaters are of special importance because these applications use large semiconductor content and have many years of product innovation ahead of them.

Many of the large Japanese consumer electronics companies have a long history of developing advanced semiconductor products. But with the right leading edge products at the right price there are still good opportunities to win long-lasting business in Japan. In general, developing a solid relationship with Japanese customers will take a long time. The upside to this is that once a relationship is established, the relationship can last for a long time.

The key to winning business in Japan is to have innovative products. Given the high cost of developing products in Japan, similar to those of United States and Europe, Japanese companies have to create high-margin products. Innovative products that target early adaptors of various electronic products can demand higher profit margins than those of commodity products. Japanese companies realize that profit margins for consumer electronics products are always under threat due to competition; therefore, they strive to deliver differentiated new products to command higher market prices.

It's often difficult to get an early indication of what a Japanese company may need from a semiconductor product. The complex organizational structure of Japanese companies is legendary. This complexity is the reason most American firms depend on their local sales and regional marketing staff to grow their businesses. Local staff can take the time to understand the decision-makers and influencers in large corporations such as NEC, Toshiba, Sony, Panasonic and many others. Even though most semiconductor products sold to Japan are via large trading companies, a marketer cannot depend on trading companies to position and promote products. It's best to establish a local presence.

One way to get the attention of a Japanese firm is to find out if there are opportunities for developing semiconductor parts jointly. By doing this, a company may need to share intellectual property with the Japanese firm. To further entice a Japanese firm to design-in a semiconductor product, a marketer could offer the Japanese consumer electronics company, such as Sony, to fabricate a portion of the final product in its own internal fab or buy it directly from the foundry used by the chip vendor.

Unconventional selling and marketing strategies such as IP or manufacturing sharing could significantly add to the complexity and cost of doing business in Japan. Yet, in some cases a marketer may need to take these aggressive measures in order to sell semiconductor products to large Japanese firms.

THE ASIA PACIFIC MARKET

Large portions of electronic products are currently manufactured in Korea, Taiwan, China, Malaysia, Thailand, and Singapore. This trend, especially manufacturing in China, will continue as more companies seek low-cost manufacturers to address ever-decreasing prices of electronic products. APAC countries are also starting to develop higher-value products that could bring more product design and innovation to the region.

KOREA MARKET

In particular, Korea is paying special attention to creating innovative electronic products that directly compete with products developed by large and established Japanese firms such as Sony. In Korea, Samsung and LG Electronics are focusing on key market segments that have been traditionally dominated by Japanese firms. These include home entertainment solutions, mobile phones, and display devices. Korean companies are also very active in Europe establishing their brands and creating markets for higher-value products, such as plasma TVs.

Doing business in Korea requires establishing a strong local sales force. Korean companies move much faster than Japanese firms in choosing semiconductor solutions, but still seek innovative products and long-term commitments to developing advanced products for key emerging markets.

Because the size of the domestic Korean market is limited, Korean companies are generally very interested in knowing how product features and cost can help them grow their businesses outside Korea. A strong selling point with Korean customers is the relationships that a chip vendor has with their potential end customers, especially in the United States and Europe.

Korean companies strive to charge a premium for their products over their Taiwanese competitors. Therefore, they tend not to chase pure commodity product markets. Korean companies also value having long-term relationships and generally pay more attention to the quality of their products than other countries in the Asia Pacific region. In the near future, a marketer may even have to broker a relationship between Korean OEMs and low-cost manufacturers in Taiwan and China, as Korean OEMs look for ways to lower their product costs.

TAIWAN MARKET

Taiwan is the center of fast design and low-cost manufacturing in Asia. Taiwan continues to expand its expertise from just manufacturing product to designing and even marketing end solutions. In addition, Taiwanese companies are starting to attack the low-end of the semiconductor market by developing their own ICs. Taiwanese companies are expert at working with U.S. semiconductor companies to lower the manufacturing cost of system products, and in the process are taking major market share away from larger OEMs in the United States, Europe, and Japan. Today, Taiwanese companies produce the bulk of building blocks of PCs sold by large PC OEMs such as Dell. In addition, Taiwan electronics firms produce a majority of the PC peripherals sold in the retail channel.

This attention to fast design cycle and low-cost manufacturing provides Taiwanese high-tech firms with a unique set of skills that cannot easily be duplicated by other countries. To do business in Taiwan, complete solutions at extremely attractive prices are required. A complete solution means exactly that: all the necessary software, system reference design, and other tools that allow a Taiwanese company to bring products to market as fast as four to six weeks. Taiwanese firms prefer to align themselves with companies that can provide them with a complete portfolio of leading-edge technologies, as they also try to bring more value to their products. Therefore, a company with a fast follower strategy will be forced to sell its products at lower prices to win over Taiwanese ODMs.

WINNING THE TAIWANESE ODMS

In addition to providing ODMs with a complete solution and low-cost semiconductor products, semiconductor companies might also have to help them create end demand for their final products. For example, say a company is selling products into the PC space, the company might need to develop relationships with PC OEMs such as Dell and HP to bring business to its Taiwanese partners. The company may even have to create demand in the retail channel to make certain ODMs can ship a large volume of their products to consumers. In the case of home networking equipment and other PC peripherals, marketers need to drive the demand with channel brands so that ODMs in Taiwan know that products will sell through the channel. Therefore, it's important to know which channel brands on the retail shelf are winners.

In general, the battle for influencing product direction is primarily with channel brands and large OEMs. By pushing products directly to channel brands, a marketer can create a conflict with larger OEMs that use his chips, since the products from Taiwan are in most cases priced well below the prices of larger OEMs located in the U.S. and Europe.

In each case, a marketer must determine who can sell more products through various channels. If products are selling directly to service providers, in the early stages of product adoption a marketer must stay with large local OEMs. The local OEMs add value to products by developing system software, end products, and marketing those products to end service providers. However, as system functionalities get integrated into a single IC (SoC), a semiconductor company will likely need to develop a complete software solution for that product. But semiconductor companies can't charge much for their system software or reference designs. Therefore, as products move to Taiwan ODMs, the value of software and system design associated with a semiconductor system solution diminishes, giving the opportunity to Taiwanese ODMs to directly compete with large established OEMs in any channel.

The Taiwanese ODM business model forces ODMs to continuously seek lower-cost IC solutions for products entering their commodity phase. Therefore, companies determined to sell large volume of products to ODMs must continuously lower product costs to be able to remain competitive. Even then, unless there is a defensible technology and patents around a semiconductor product, the product may still be forced out of the market by a lower-cost semiconductor product designed and produced in Taiwan. Taiwanese firms have been very active in developing IC solutions in recent years, as they move to protect their revenue from the lower-cost manufacturers in China. This means that, in the next five to ten years, many of the current low-end semiconductor products will be designed and manufactured by Taiwanese firms. The manufacturing of such semiconductor products could also eventually find their way to

China, as the Chinese semiconductor foundries increase their share of the worldwide semiconductor IC market.

As was previously noted, channel brands in the United States and Europe get the majority of their products directly from Taiwanese ODMs. The channel brands, however, often depend on semiconductor companies to introduce them to the right ODMs in Taiwan and other parts of Asia. A marketer should strive to get the tier-one ODMs in Taiwan to design in his products, but may have to settle for second-tier ODMs. The second-tier ODMs may not have the financial and technical strengths to provide the channel brands with the right products at the right price.

In some cases, an IC product ramp may be limited by the ability of ODMs to design and manufacture the system solution. If this happens, a marketer might have to embark on the expensive, but necessary strategy of giving ODMs in Taiwan a complete solution comprised of hardware, software, and manufactureable reference designs. This strategy works when product volume rapidly grows to justify the financial burden of developing a complete solution.

Providing a complete solution to every ODM becomes problematic when the ODMs decide to differentiate their products. Each will require unique support for a unique set of features and capabilities. ODMs will often seek differentiation when selling to the retail channel, where product functionalities do not depend on a set of standards defined by service providers. In the end, it's up to marketers to decide if supporting multiple system solution is feasible for a given market segment. Marketers have to choose and support those system solutions most likely to achieve revenues—those backed by big-name channel brands.

CHINA MARKET

Beyond working with Taiwanese ODMs, marketers must start working closer with ODMs in China. Chinese manufacturers continue to grow their share of manufacturing of electronic devices for export to various parts of the world, especially the United States. The domestic Chinese electronic market is also large and in most cases undeveloped. U.S., European, Japanese, and even Korean firms are spending lots of marketing dollars to establish their brands in the mind of Chinese consumers. In addition, through various joint ventures, large outside firms have established partnerships or design centers to produce products specifically designed for the China market.

Developing the Chinese market requires long-term planning and lots of patience. The fact is China is not currently a large market for various semiconductor solutions. But, China has huge potential. Semiconductor companies must have a plan to tap into the huge pool of highly skilled engineers and programmers to develop products not only for the China market, but also for markets outside of China. Taiwanese firms have already been active in setting up factories in China to lower their product costs. In the process, they are teaching the Chinese how to develop highly complex and low-cost electronic products.

To develop long-lasting relationships in China, it is a good idea to work closely with universities that are training the largest and brightest number of engineering and computer science students. It may be necessary for semiconductor companies to provide development tools and financial help to top universities in China. This will encourage faculty and students of those universities to use a company's semiconductor product and utilize them in their projects. These students are tomorrow's engineers, decision-makers, and leaders in Chinese firms.

Semiconductor companies also should analyze and understand the role of the Chinese government in every aspect of technology development in China. The Chinese government heavily influences technology standards utilized there; be ready to address those standards. A marketing effort can influence the direction of standards by developing key relationships with various government offices. All of these would require having a local presence in either China or Taiwan that can fluently communicate with Chinese engineers, professors, and government officials in their local languages.

By moving product manufacturing to China, the Taiwanese free up their resources to develop more complex ICs. In time, Taiwanese design firms will transfer the manufacturing of their semiconductor products to China, further lowering the selling price of semiconductor ICs developed in Taiwan. This could have a long-term and negative impact on U.S. semiconductor companies that still produce commodity products. During the past two decades, the United States and Europe have lost most of their wafer fabrication business to Asia, especially to Taiwan. If the U.S., European, and Japanese firms do not continue to innovate, they could end up losing the bulk of their semiconductor product revenue to Taiwanese and, in the long-term, to Chinese firms.

SUMMARY

The United States, Europe, Japan, and Asia Pacific are the major consumers of semiconductor products. Each market region has its own sets of dynamics and business rules. Depending on a product line, a marketer will have to play in at least two, if not all, of these regions to achieve revenue success. The growing dominance of Asia as the center of consumer electronics manufacturing will likely mean that, regardless of the target market, Asia ODMs and OEMs will play a key role in success or failure of any product.

forecasting ›

A difficult challenge for marketing is providing consistent and accurate forecasts of revenue and margin for product lines. Forecasting and delivering on revenue and margin is really the only solid metric that marketing can use to gauge the success of a new product. Marketing would not be very challenging if it were only about identifying target markets, defining products and program management, without having to deliver actual revenue. Determining whether a product line is living up to its promise is a question of measuring actual revenues against forecast.

To add to the marketer's challenge, forecasting is not an exact science. Unlike engineering design where the use of mathematics and simulation tools can help predict the behavior of systems, there are few simulation tools to help do an accurate forecast. There is no "forecast equation." Economists try to use econometrics to forecast economic growth and other economic parameters and routinely fail to be able to predict future outcomes.

The problem is that too many dynamic variables can affect the accuracy of a forecast. Yet, forecast accuracy does improve with knowledge and experience with a product line. Knowledge of a product, the target market, the value chain, competition, and the inter-dependencies of a product line to other factors (such as PC demand or telecommunications capital expenditures) are all necessary for effective forecasting.

TOP-DOWN FORECAST

Even though forecasting is something that most semiconductor marketers would like to avoid, it's the guiding light for running their product lines. A marketer who is conservative and tends to give low revenue targets may not get the necessary resources needed to grow his business. On the other hand, consistently forecasting optimistically can lead to loss of credibility, since actual numbers will be less than estimates.

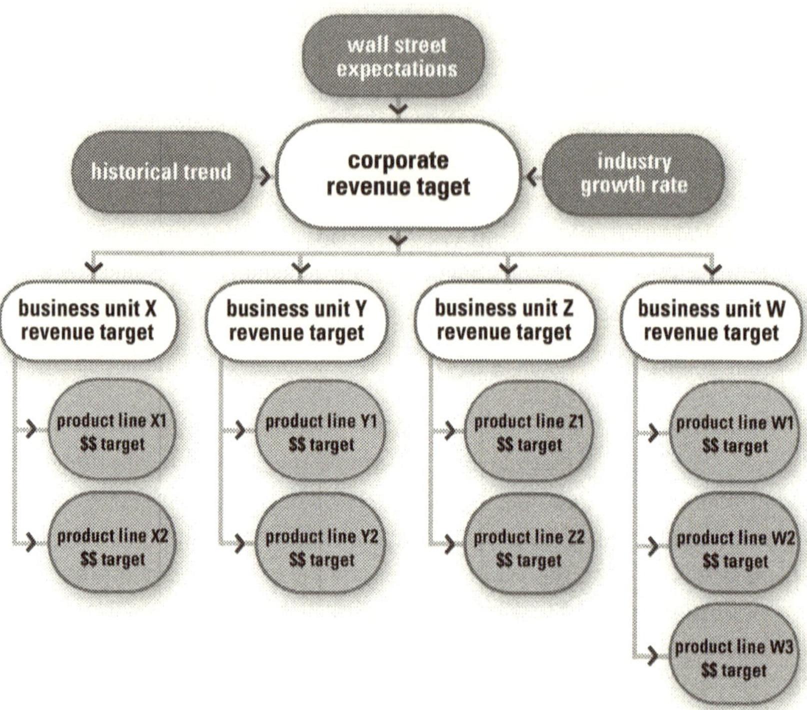

Generally, forecasts are developed during the initial business justification of products. These forecasts are derived from making certain assumptions about market size and the segment of that market a new product will address. These initial forecasts get revised as products come to market and start generating revenue. Once a product ramps, the forecast will be bounded by some expectations based on actual revenue and a host of other dynamics. It is at this point that a product will enter its normal forecasting phase, which starts with what is called a "top-line" or "top-down" forecast.

Senior management of public corporations must give revenue guidance to either a board of directors and/or financial analysts covering the company stock. Typically, a few months before the start of a fiscal year, senior management together with sales and finance will determine a revenue goal for the whole year. This target number is based on a estimated summation of revenue target numbers for each business unit. Each business unit justifies the revenue target across individual product lines.

Of course, the top-down forecast process is usually iterative. Management pushes business units and product line managers to commit to higher revenue numbers. Sales and marketing, at the same time, try to modulate the numbers based on their understanding of their customers and products. It's vital for marketing to participate in the development of these top-down revenue targets.

If a marketer feels that the top-down target is unachievable, he must communicate that to his senior management. The marketer must also indicate when the top-down revenue target is too low. Either way, it is a marketer's job to provide the most accurate revenue possible to help management plan properly in the development of the top-line number.

Senior management often arrives at the top-down revenue number by looking at historical data and examining market trends that influence product lines. In the process, senior management may decide to shift investments more toward product lines that have brighter futures. The process results in an overall portfolio positioning of a company. This shifting of financial and human rosources away or toward a product line is a major factor influencing how successfully a marketer can develop achievable revenue goals.

In doing a top-down forecast, managers and marketers should not underestimate the threat of competition. Even a lone wolf in a market segment should assume the worst. High margins and volumes tend to attract competitors, unless there is some kind of monopoly that prevents them. Marketers should take competition into account by coming up with best, typical, and worst-case scenarios to realistically model revenue targets. Doing a revenue-sensitivity analysis is a good practice in highly competitive and dynamic product markets.

As we noted, coming up with top-line revenue forecast is a shared iterative process. But, if management took no interest in a particular product, the marketer in charge of that product should still develop a top-line target or goal for the year. Nothing is achieved without setting a goal. Top-line revenue targets are goals for companies, business units and individual products. They are important part of the forecasting process. Once a target goal is set, a detailed customer by customer forecasting work needs to happen, this is referred to the bottom-up forecast.

BOTTOM-UP FORECAST

The problem with doing a top-down forecast is that it does not specifically show where the revenue will be coming from. For this, marketers rely on the bottom-up forecast. A bottom-up forecast lists revenue per customer over time. In a way, the bottom-up forecast is a marketer's life-blood. It shows exactly how the money is or should be coming in. It must be constantly updated to ensure that revenue goals are on track with the top-down forecast, as well as to monitor potential problems and/or up-sides to incoming revenue. Therefore, this forecast should be monitored and updated on a regular basis (at least monthly).

To develop the bottom-up forecast, a marketer must work with sales to determine how much product each customer is planning to purchase over a specified time period. Salespeople should know the status of product design-ins for each account and the time frame when they expect to get orders from their customers. Sales insight, together with a marketer's knowledge, will help determine how much success customers will have in their end-markets. Working closely with sales also ensures that sales and marketing are "on the same page" when it comes to forecasts, which is crucial to doing accurate bottom-up forecasts.

Relying completely on customer forecasts is not always the most accurate way to judge future product revenue. It is common for customers to initially place large orders for products and then to subsequently cancel the orders before or after production ramp. In some cases, customers simply find that there is little or no demand for their products.

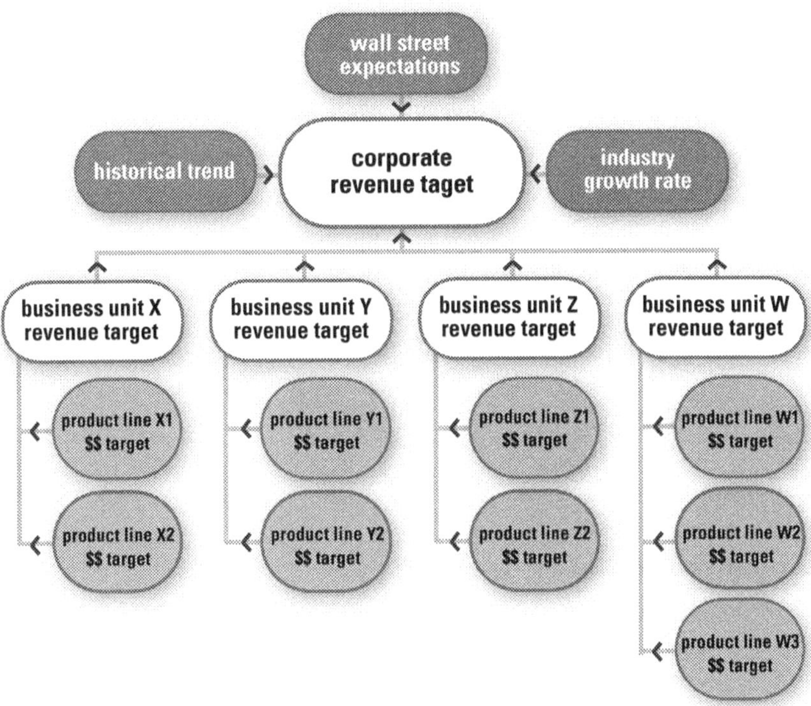

Therefore, it is critical for marketing to pay special attention to information about customers' end-markets. For example, say a marketer is faced with a sudden increase in demand for his graphics-controller ICs for PCs from a particular customer. The marketer should use sales contacts, his knowledge, and any third-party contacts to determine if his customer has won a major business at any of the PC OEMs. If the order cannot be justified, a marketer must remain wary of cancellation.

The best way to devise a bottom-up forecast is to create a spreadsheet listing the names of customers and what their demands are each month, extending at least 12 to 18 months into the future. On the same spreadsheet, marketers should specify the ASPs for each customer for past and future orders. The ASP multiplied with the forecast volume determines revenue forecasts for each customer. Each customer forecast is then added to provide total revenue projections

PRODUCT TRANSITIONS AND SALES CYCLES

When doing a bottom-up forecast, marketers must pay special attention to when and how to transition a customer from an old to new generation of products. Badly planned switchovers could lead to costly oversupply and undersupply of both old and new products. In most cases, customers do not necessarily move to a new product at the speed marketers would like them to. Marketers must understand customers, design cycles, and channel inventories to determine how quickly platforms can switch to new products.

In moving customers to a new product line, marketers have to pay special attention to product readiness. Marketers often push new products too fast without having enough manufacturing data to see if new products yield properly, and that the quality of the products can be assured. It's better to transfer only a few customers to a new product to make certain no problems crop up in limited production and deployment. The transfer of all customers should take place once there is confidence in production of the new product.

One factor that determines transition time to a new product is the inventory level of the old product. If there is large inventory of the older product, even if it costs more than the new product, it represents sunk cost. In comparison, the new product must be bought with today's dollars. Therefore, marketing will want to deplete the inventory of the old product before switching customers over to the new product.

However, some situations dictate that switchover to a new product happens as fast as possible. For example, if a new product represents a technological lead over the competition, then moving customers over quickly is an advantage. One should always strive to close the window of opportunity on competitors to design-in their products.

REVENUE FORECAST UPDATE

Given that most companies have to provide the financial community with quarterly results, short-term forecasts must be as accurate as possible. Short-term forecasts typically are one quarter out. To increase quarterly accuracy, marketing at a minimum should update forecasts every month if not every week. Weekly updates will help prepare a marketer for his monthly revenue meetings with management and planning.

Regular forecasting also ensures that any problems can be caught early. Regular updating requires continuous feedback from field salespeople. After all, salespeople have to chase orders and make sure money flows into the corporation.

If salespeople are not providing marketing with information, marketing should travel to target regions each quarter to generate finalized quarterly numbers. This may be a great way to help salespeople close business and ask customers first hand what their perspectives are on market opportunities.

MARKETING INFORMATION SYSTEM

A marketing information system is any set of tools that allows marketing to track customer design activities, product demand, revenue forecast, average selling prices, and product development status. Most semiconductor companies do not have a comprehensive information system that provides marketing with the state of their business. Instead, companies typically have a large number of support staff that tracks various aspects of product line performance using simple spreadsheet files. These old methods of managing a business might work for a little while, but may not be adequate when the business becomes larger and more complex. For example, a marketer may not be notified when his supply-management group places wafer orders for his products. But the marketer is the one who has to answer to senior management about inventory levels and why too many wafers were built.

Ideally, marketers should have the same tools as salespeople to track customer design-wins and unit demand. If a company creates specific marketing tools in isolation without having them tied to other enterprise tools, then the system will require too many full-time people to maintain.

SURVIVING FORECASTING

A product line with a dominant market position that is at least several product generations old is easier to forecast than a brand new product with little or no track record. Either way, management uses revenues and margin as a scorecard for a products success. Product success or failure reflects directly on the marketer in charge of the product. A marketer whose products consistently fall short of forecast can be singled out for criticism or worse.

Of course, a marketer is not the sole person controlling all the elements that can affect the success or failure of a product line. There are always elements outside of marketing's control, such as macro-level product trends, development team execution, and competition. Yet, as a marketer, too many under performing products can be detrimental to a career. Or at least, the lack of product success will not help a career.

Even during the darkest moments of a revenue cycle, when the revenue chips are down, a marketer must stay positive about his product line. He should never go around blaming others for the revenue shortfall. Instead marketers should devise an action plan to correct the situation. There are situations (for example, when a competitor decides to exit or enter the market) that can cause the dynamics of a market segment to change and prices to drop quickly. The price drops could have a short- or long-term effect on the dynamics of a product market, but a marketer must make sure management is aware of that event ahead of time.

There are also cases when a development team simply cannot execute to a schedule. Marketers must regularly highlight to product development what effects product delays have on revenue forecasts. Marketers should stay focused on the objectives and have the courage to recommend cancellation of products if they are running well behind schedule.

Marketers must manage their product lines in all circumstances—even dire circumstances.

In the case where ASP declines are the cause of revenue shortfall, marketers must determine whether cost reductions are necessary to maintain margins. In the semiconductor business, prices rarely increase, except in the memory business. Therefore, a marketer may have to consider the feasibility of cost-reducing a product to improve profit margin. If prices drop and there are no ways to reduce cost, a marketer should consider exiting a business. Revenue shortfalls, unless due to short-term macro-level developments, are serious indicators of the health of a product line and its target markets. Revenue shortfall also could be due to competitors gaining market share because they have superior products and cost advantages.

Marketing must take revenue shortfalls seriously and work through the issues to understand the root causes of the problem. Marketers get paid to manage the business during both the good and the bad times. Marketing must keep a steady hand on the tiller, ride the events, and stay firm on strategic plans.

SUMMARY

Too often marketers do not take forecasting very seriously. They look at their numbers once a month, and sometimes for only ten minutes before a forecast meeting with their management. There is no way a marketer can understand his business by looking at the forecast ten minutes a month.

Marketing must look at the bottom-up forecast at least weekly and compare it to top-down numbers regularly. How a product is delivering to forecast is a barometer of its health. Marketers must forecast regularly to ensure that the business is on track and must also be ready to take action when products are going off track. Forecasting is like exercising — you need to do it regularly in order to get the maximum benefits.

sustaining legacy product lines ›

Established semiconductor companies typically have at least a few mature product lines that bring in most of the revenue and profit margin. Companies refer to these product lines as "legacy," "core," or "foundation" businesses. One aspect of these legacy businesses is that they are stable "money makers" that dominate their market segments.

Throughout this book most of our comments have centered on what it takes to bring a product to market. But marketing is also about the care and maintenance of legacy product lines. This chapter will delve into more detail about how to manage these core businesses.

VALUING LEGACY PRODUCT LINES

Legacy product lines do not typically grow much, if at all, and usually suffer price erosions as competitors try to enter the market. In addition, due to the maturity of the functionality of legacy products, using product differentiation to extract value or fight competition becomes more difficult. These factors make it challenging to manage legacy product lines. However, it is usually in a company's best interest to extend the life of a legacy product line in order to bring in much-needed revenue and margin, to fund the development of newer products necessary for future growth.

Semiconductor companies typically do not value their legacy product lines, especially during boom years. Often, both management and employees are seduced by the promise and glory associated with new product lines. In addition, since legacy product lines are often not technically interesting, key engineers are happy to lay aside these "old dogs." Most people in semiconductor companies want to be associated with newer product lines that are in the spotlight. The financial community helps fuel this tendency by pushing company executives to enter newer product spaces seen as cutting edge. This means that marketers responsible for managing legacy product lines are often in a thankless job with little top-management exposure. In addition, it may be difficult to secure enough resources to manage and maintain the legacy product lines.

In spite of these challenges, a semiconductor marketer must promote and highlight the importance of a legacy product line to senior management to get the attention it deserves. Most legacy product lines do not get obsolete as fast as planners predict. Hence, one can usually squeeze much more revenue and margin dollars from a legacy product line than many expect. If a product can maintain a strong position until the end of a market lifecycle, fewer competitors will bother to attack. This is not surprising since potential competitors will realize it is not worth fighting the incumbency in an ever-shrinking market. Likewise, venture capitalists are not known to finance start-ups that plan to enter mature markets.

Illustration 27: product profit lifecycle

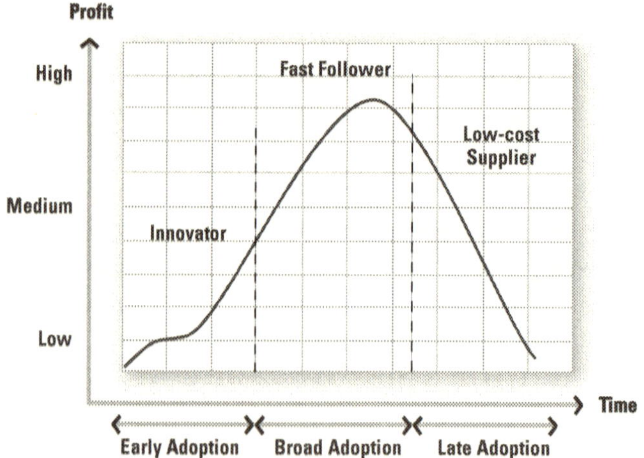

To successfully manage a legacy product line, a marketer must be able to manage cost and quality of mature products with minimal investment. In addition, a marketer must put a plan together to service customers at a much higher level, and turn ordinary customer service into a source of sustainable competitive advantage. Managing a legacy product line offers a great opportunity to learn the true art of selling and marketing without having highly featured differentiated products. As was previously noted, if a product has highly differentiated features, is very competitively priced, and supported well, the product sells itself. Legacy products, on the other hand, need intelligent marketers and salespeople to maintain their success.

MANAGING SUPPLY AND DEMAND

Managing supply and demand is important for any semiconductor product line. But it's crucial for legacy product lines. Typically, it is easier to forecast supply and demand for legacy product lines since the business is established. In addition, after many years in the business, a lot of useful data showing "actual" versus "predicted" product shipments exists. In most cases, there may be many people in a company associated with a legacy product line over the years. These people have lots of insight into the dynamics of the market and competitive nature of the business. A marketer can seek their input when faced with unusual demand patterns.

Assessing an accurate demand for legacy product lines also lowers the risk of building wasteful inventory. Unlike newer product lines, legacy product lines may go through significant demand changes rapidly, if a major customer decides to switch to a lower-cost product or integrate the product out. As switching costs are lower for mature products, the customers' ability to switch to a competitive solution is increased. Needless to say, the loss of a big customer will make a significant effect on demand.

Another benefit of managing demand is always having adequate inventory so that customers get products on time. Customers always remember a supplier's track record and will retaliate as soon as they get a chance. Therefore, a marketer must carefully manage product line supply and demand.

As part of establishing a system to manage supply and demand properly, marketing must communicate regularly (once a week) with operations people to track product yields, returns, and wafer foundry starts and capacity. Marketers must ensure that operations people are familiar with the business and the top ten customers. This will allow operations to make better judgments if they foresee wafer delays and yield issues. Marketers learn with experience that operation organizations are usually understaffed and overworked. To make sure that operations people remain vigilant, marketers have to take the time to develop good working relations with them.

UNDERSTANDING PRODUCTION FLOW

To better prepare for managing all supply related issues, marketing must acquire in-depth knowledge of production flow and processes. The best way to do this is to actually spend some time on a production floor to observe firsthand what it takes to manufacture products. Also important is to understand what causes products not to yield and what controls production people have put in place to test and ship products. Marketers should learn how their company's operations people manage the supply chain for products as well as all aspects of the quality system and policies. At the end of the day, this knowledge helps marketing manage their businesses because, as problems come up, they can proactively make suggestions and find solutions.

For legacy product lines, producing high quality products is a huge source of sustainable competitive advantage. If a company is known to have higher quality products, then there is a better chance of keeping current customers and even getting new customers, assuming products have the same features and are priced competitively. In some cases, marketers must work closely with suppliers and customers to make sure everyone involved in the supply chain can easily monitor each other's quality trends.

As part of implementing a world-class quality control system, it is necessary to monitor production parameters and test data. The only way to know something is going wrong with a product's yield and quality, or has gone wrong, is to monitor the product's performance, yield, and quality. In many companies, data analysis and monitoring is an afterthought. This can eventually lead to the inability to give customers information that they need when yield and quality issues affect their end product performance and increase field failures.

All marketers will experience major customer yield and quality issues during a product's life. This is typical of the industry. It's not a question of if it's going to happen, it's a question of when. When it does happen, a marketer's ability to quickly analyze production data and arrive at corrective actions makes a huge difference in the rapid resolution to problems. Marketers need to be paranoid—and always have contingency plans in place. Marketers also should create mock disasters that give them the chance to think about how to deal with emergency situations. For example, assume 75% of a product starts failing in the field. How would you handle the issue? What tools do you have at hand to determine the root cause of the failure? This is very similar to having a fire drill. Marketing's ability to manage through crisis is key to managing legacy product lines successfully.

DEALING WITH RMAs

As part of running any semiconductor product line, marketers have to deal with major product return issues or "RMAs." RMA actually stands for "Return Material Authorization," but the acronym is used to describe products returned by customers. Many factors can affect the quality of products shipped to customers. These include not having good wafer process monitoring, low test coverage, no outgoing quality tests, no life-test and burn-in, and lacking a systematic failure analysis on all product returns. Whatever the reasons, the end result is the same. Customers will end up receiving products that could cause their systems to fail during production tests, or worst case, force them to deal with field failures.

Larger customers will require a guarantee that vendors perform failure analysis on parts within specific guidelines. In some cases, companies such as Dell may even require a supplier to monitor and resolve product yield and quality issues within hours. A marketing manager responsible for legacy product lines has to ensure that operations people are aware and ready to deal with such requests.

When receiving a large number of returns, marketing together with quality and product test groups must determine first how widespread the problem is. For that, product yield and quality data need to be collected. Having access to a large amount of data, and the ability to analyze it quickly, will give customers a view into the severity of the problem. Without having adequate data, customers will assume that their entire inventory has the problem. In addition, customers may even assume that all of their systems in the field could have the problem.

Customers know semiconductor products are not perfect and will go through yield and quality issues. Customers, however, will remember how quickly a vendor was able to solve a problem. They also may have to be compensated for all their field and product failures. Therefore, the financial consequences of major field returns could cost more than simply replacing defective parts. Quick resolution of RMAs by bounding the problem and finding its root cause will help limit the potential damage incurred and avoid panic by all involved.

COST REDUCTION EFFORTS

The task of sustaining the revenue of an established product line includes the primary task of looking for ways to reduce products' costs. Once a semiconductor product market matures, many companies have the technology to enter those markets. If the market dynamics are favorable for outsiders, new competitors will enter the market rapidly.

To prevent new competitors from entering a mature product market segment and to protect market share, marketers must establish an effective cost-reduction plan especially for high-volume products. The objective of this plan is to constantly look for ways to lower product costs. A marketer could start by examining how much overhead and manufacturing costs are allocated to a product. Such an investigation may reveal that an unfair share of manufacturing overhead costs is being applied to the high-volume legacy product line.

If the manufacturing department refuses to lower product overhead costs, a marketer must look to other ways to reduce the overhead costs. One way is to outsource test or wafer fabrication of products, assuming those are normally handled in house. In addition, a marketer must examine the complete cost model of a product line to determine which cost items can be reduced. In some cases, if the internal wafer-fabrication facilities are not running at full capacity, a product may get hit by more of the overhead cost associated with using the internal wafer foundries. If that happens, it is necessary to challenge the finance people to see if they will associate only the directly relevant costs.

Marketers should also examine packaging and assembly costs. Once again, a marketer must be wary of getting charged with excess overhead charges when using internal facilities to package and assemble products. In addition to packaging and assembly costs, it is also important to examine test costs and the amount of time required to test products. Every second needed to test a device could add up to be a significant cost for a high-volume product line. Marketers must challenge test and product engineers to continuously reduce the test time and increase the yield of products.

Besides test costs, the number of test insertions in a typical test flow can increase costs. For example, production people may be testing products at cold, room, and hot temperatures to make sure the products meet published specifications. It may be possible to get away with only a single insertion test, which results in significant test cost reduction.

After reviewing all of the test costs, it is important to make sure the costs applied to a product accurately reflects the actual yield of the products. It may be that a product cost model assumes the final test of a product to be 90% when in reality the actual yield is 95%. In this case, the cost applied to that product ignores the additional 5% yield. Marketers should establish a process where actual yields are continuously applied to the product cost model. It's the job of product engineering groups to maintain and improve product yield. The higher the yield is, the lower the cost of products.

After completing the easy cost-reduction tasks, marketers must review each product's die size and process technologies to determine if it makes sense to shrink the die size of high-volume products. If the product design can easily be ported to a new process geometry, marketing should model the cost savings and quickly determine if a port to the new technology is justified.

If the product volume of a system solution is not high, marketing may want to limit cost-reduction activities. Instead, it may be more important to spend more time developing new products that can give the feature differentiations customers may value.

SUMMARY

Managing legacy product lines well is a valuable art, since everyone appreciates a stable income. While it is true that people who manage established product lines sometimes do not get the attention and recognition they deserve, it is a great way for a marketer to hone his business skills. In fact, successfully managing a core product line is harder in some ways than launching a new product line. When a marketer cannot make money running a legacy business, it looks like he drove a good business into the ground. This is not true with a new product line, where failure may be harder to measure.

The key to managing an established product line is to stay vigilant. Vigilance is the message through this whole chapter. It is critical to understand the challenges associated with a legacy product line as acquire all the relevant knowledge needed to run the product line. Understanding production and managing supply must become a priority. Knowing what can affect yields is also important. Managing costs is a full time job necessary to milk profitability. Vigilance touched with paranoia coupled with a ruthless drive to understand the business, are the qualities necessary for managing a legacy product line well.

understanding marketing organizations >

How a manager organizes a marketing group depends on factors related to the target market as well as the product. The two most common organizational structures are vertical and functional. In a vertical organization, each marketer is in charge of an entire product category covering all aspects of new product development and product lifecycle management. In a functional organization, tasks such as customer marketing, product marketing and technical marketing are divided among a set of marketers.

illustration 28: typical marketing organization

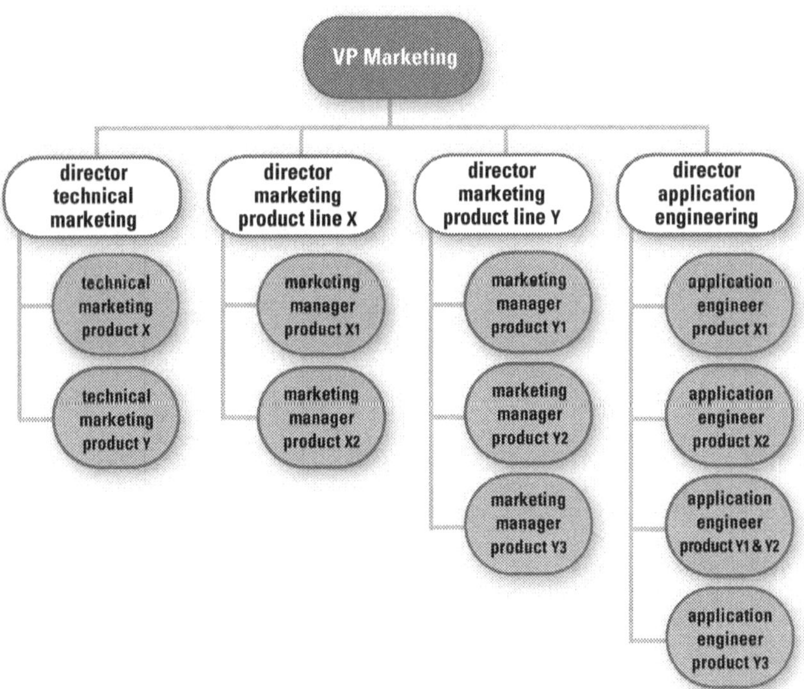

VERTICAL MARKETING ORGANIZATION

A vertical marketing organization is perhaps the most straightforward. It is easy to associate a product with one "go to person". The marketer is responsible for defining new products, working with development groups to get the products designed, introducing products, performing all customer-marketing tasks, managing product pricing, and overseeing all supply and demand issues. However, in a vertical marketing organization, depending on the complexity of a product, a single person might not have the bandwidth to perform all the marketing tasks well.

Product-marketing tasks and customer-marketing tasks require two different skill sets. In most cases, people tend to gravitate toward one or the other. The product-marketing people are more internally oriented and the customer-marketing people are more externally oriented. However, in a vertical organization, one is expected to have a complete and comprehensive set of skills.

illustration 29: vertical marketing responsibilities

If a market is made up of a small set of customers and a road map that needs only one or two products, a vertical marketing organization will suffice. A vertical organization is also appropriate in a start-up company developing its first few products. Specific market dynamics or technology may also favor a vertical marketing organization structure.

In some cases, vertical marketing organizations have advantages over functional organizations. A single person in charge of all activities is the focal point to whom everyone can turn. Product development people, salespeople, and customers will never be confused about who is responsible for various product issues and problems. A vertical organization also works well when there are major problems to solve or crises to respond to. The ownership of the product line is clearly understood.

However, a vertical organization also has its drawbacks. At a minimum there are logistical complications when all power is vested in one person. For example, when a marketing manager is traveling, issues may arise and resolution to problems may have to wait until the marketer returns. In addition, customers might not get the attention they need from the marketing manager, when the marketer is busy defining or launching a new product. Limited marketing bandwidth may also lead to limited strategic thinking. So, even though from an accountability standpoint, a vertical marketing group can be easier to manage and measure, a vertical organization may place too much responsibility on one person.

FUNCTIONAL MARKETING ORGANIZATION

Many of the vertical marketing organizational challenges are solved with a horizontal or functional marketing organization structure, where various marketing people in the group take responsibility for a certain set of tasks. For example, in a functional organization, customer and strategic marketing activities can be separated from product-marketing activities. A functional marketing organization allows marketers with the appropriate skill sets to be placed in each of these groups.

In a functional marketing organization, a marketing vice president or director has to make sure roles and responsibilities of the marketers in the group are clearly defined and communicated. Otherwise, there is a risk of marketers not knowing what their jobs are. This could lead to confusion in the minds of engineers, salespeople, and customers.

However, a well run functional marketing organization provides the benefit of allowing the organization to manage a larger number of products. It also allows the marketers to perform each set of product marketing and program management activities better.

A functional marketing group is especially appropriate for companies pursuing large system-level products that have significant software and hardware components and integration challenges. Systems solutions by definition require a large amount of customer support.

illustration 30: functional marketing responsibilities

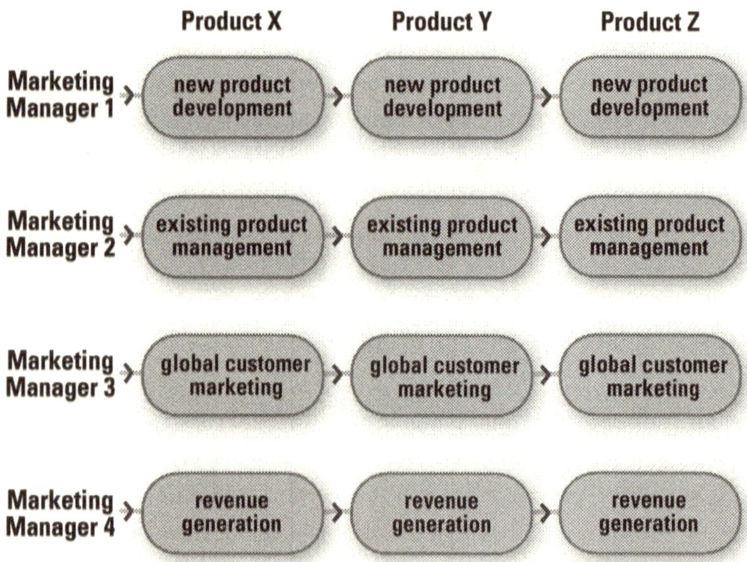

Functional groups are also better equipped at dealing with multiple industry segments. For example, let's assume PC OEMs, like Dell and Hewlett-Packard (HP), and consumer electronic companies such as Sony can use a particular product. Depending on the extent of each customer group's activity and demand, a marketing vice president could assign a marketer to manage the PC OEMs, and another marketer to cover the consumer electronics market segments. However, a single marketer could still be in charge of developing requirements and launching products that would cover both industry market segments.

One issue with a functional marketing organization concerns the limited role that each marketer has. This means that each marketer has a boundary around what he can and cannot do. Some marketers have a problem with functional marketing groups, because it does not allow them to have total ownership of an entire product line or product category. This is often an issue for experienced marketers or marketers who have worked in small companies and have been in charge of an entire set of marketing activities.

OPTIMIZING MARKETING ORGANIZATION

A manager who is responsible for a marketing organization will have to continuously consider ways to optimize his group. Part of this is to determine the skills and the career goals of his team members. A manager may find that what his marketers want to do does not best fit their skills and background.

A manager must evaluate his marketing team members based on their strengths and weaknesses. This determination is easy — sometimes. For example, some people gravitate toward product marketing. They tend to be introverted and are successful at defining new products and getting those products through engineering. These people may have little interest in daily issues like revenue forecasts, dealing with salespeople, or meeting customers regularly. Others may have the opposite set of skills — extroverted and enjoy meeting customers and dealing with salespeople regularly.

However, determining marketers' strengths and weaknesses is not always clear-cut. Many marketers do not spend much time where their true strengths lie. In addition, there are many instances where marketers with a certain skill set and bias actually want to do the opposite of what they are good at. Many marketers also confuse their short-term activities with their real long-term responsibilities, and therefore they neglect to prepare themselves for what is needed from them to best contribute to the success of a product line.

It is a group manager's responsibility to coach his people so they can observe their own behavior, in order to understand how they can best contribute to the success of a product line. The following definitions will help a group manager understand the role of various marketers better and help in establishing a marketing organization. The following descriptions and roles will also help marketers consider their own skills and interests.

MARKETERS' INSTINCTS: FARMERS VERSUS HUNTERS

One method for placing marketers in the right role is to determine whether they fall into the category of "farmer" or "hunter." A farmer is someone who enjoys harvesting the fruits of an established product line. This type of position requires tending to products, maintaining market share, maintaining profit margin, and watching the competition to make sure they are kept at bay. A big part of a farmer's job is to monitor product costs and product breadth. Farmers are good at working closely with manufacturing and operations. They do not mind doing the same tasks weekly and daily. They are energized by the idea of defending an existing position and maintaining market share and profit margin. Farmers also best suited to run legacy product lines.

Hunters also exist in most organizations. These marketers are interested in uncovering new markets and bringing new products to market. Members of this group feel comfortable in uncharted territory and enjoy entrepreneurial challenges. In contrast to farmers, hunters get bored with doing the same tasks day in day out and would have a hard time sustaining an existing product line.

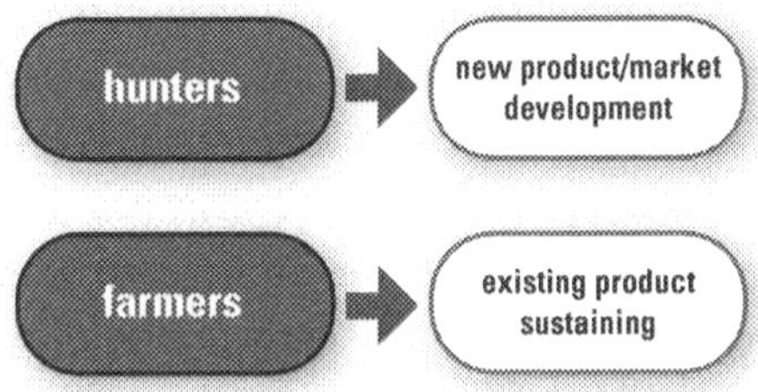

ROLE OF STRATEGIC MARKETING

Depending on the size of a marketing organization, there may be room for one or more marketers who focus on long-term projects and partnerships. In a sense, these marketers focus on the strategic direction of a product line and not so much on the day-to-day tasks of running a business. In some cases, strategic marketing people like to create new businesses and do not want to get daily or weekly directions from their managers.

People who are responsible for strategic marketing have a vision of where their respective industry will be in the next five to ten years. They can see their way through short-term crises and provide a focus on the long-term needs of product linos. In doing so, strategic-marketing people work closely with internal research and development staff to assess what is needed to develop products targeting markets three or more years down the road. If needed, strategic marketing people can also search for third-party suppliers of technologies and develop various licensing agreements so that a product line's vision can be realized by product marketing staff.

Strategic-marketing people also tend to stay close to standards committees and influence the direction of specific industry segments. It's extremely important that a strategic-marketing person develops a name for himself in the industry through various forums and organizations. He should command respect from other companies and industry veterans. The strategic-marketing role is the ideal job for people who have lots of industry contacts and can develop personal relationships with lead customers and chief technology officers of target companies.

ROLE OF PRODUCT MARKETING

Product-marketing people are primarily responsible for bringing new products to market and managing product lifecycles. These people should have a good understanding of the chips they are responsible for. Product-marketing people work closely with designers and other development engineers to determine the resources and time required to bring new products to market. It's crucial that product-marketing people gain the respect of key designers and show that they understand the trade-offs that could affect the cost, performance, and development time of new products.

This means, it's important that product-marketing people be given adequate time to understand the functionality of various chips, and to show customers as well as design engineers that they have engineering as well as marketing skills. Product-marketing people may have to spend as much as 80% of time in the office, while spending the rest with customers, to do their job well. It's also the job of product-marketing people to evaluate key attributes of competitors' products and position their products against competitive offerings. A product-marketing person must continuously update both design and sales organizations regarding the competitive nature of the industry, and show them he has the right product definition to allow the company to effectively compete in a given market segment.

A product-marketing person is also responsible for managing product demand and supply, and must work closely with production people to assure the right amount of products are built and shipped. In addition to managing supply and demand of a product line, product-marketing people are also responsible for continuously finding ways to lower the cost of products. It's a fact that without having a strategy for lowering cost or increasing functionality, the lifecycle of a product may be limited.

High-level pricing strategy and promotional activities are also usually the responsibility of a product-marketing person. As we noted previously, it's often easy to forget about the need to promote products after they have been designed and sampled by customers. Some marketers and engineers believe that, by having a great product, one does not need to promote it or aggressively price it, believing that a good product sells itself. This is wishful thinking and the mark of an engineering-driven organization.

ROLE OF CUSTOMER MARKETING

It's difficult to cover all aspects of product development and simultaneously manage customers. The solution to this problem, assuming there are a large number of products in a product line, is to establish a customer-marketing role. Customer-marketing people must have or develop more selling skills, and must be able to communicate effectively the attributes of products to customers. By having a focused customer-marketing group within a marketing organization, customers will have a primary contact they can depend on to resolve all their technical and business-related issues. In addition, sales people can use the customer-marketing people as their liaison to engineering organizations.

It may also be necessary to have a customer-marketing person covering an industry segment. For example, if a company is selling various products to PC-OEMs, then it might be a good idea to have a marketing person work with salespeople to cover all PC-OEMs. By doing this, this person can develop long-term relationships with key decision-makers in the PC-OEM market segment. Also, if a company is selling products to service providers such as cable companies, satellite broadcast companies, or telephone companies, it is probably beneficial by having a single customer-marketing person covering several service providers—rather than having multiple marketing people, each responsible for a product, covering the same customer.

It takes time to truly understand the nature of each market sub-segments and the behavior of key decision-makers within a large organization, such as a PC-OEM. So, by having a focused effort and one person covering each market sub-segment, that person would have the time needed to understand what it takes to sell products into a specific market. Also, a sales organization can then depend on that person to provide all the key materials needed to communicate the value of a product line to customers of given sub-segments.

It's difficult for a marketer to cover all aspects of product- and customer-marketing. It's also difficult for a customer-marketing person to cover various regions of the world. For example, if products ship primarily to Asia, then an organization should have one customer-marketing person covering that region. For international customers, it may be a good idea to hire marketing people with language skills necessary to handle key geographic regions. Customers like to deal with people, both marketing and sales, who understand their language and culture. Even though English is the primary language of business, many business relationships are developed and maintained outside the boundaries of the formal business world.

Customer-marketing people also manage design wins and ensure that those wins turn to volume business. Typically, a customer-marketing experience is hectic, with frequent interruptions from salespeople and customers. The term "fighting fires" probably came from a group of semiconductor customer-marketing people. In addition to numerous interruptions, customer-marketing people are also on the road frequently, visiting customers and sales groups. Often these trips are made on very short notice. Customer-marketing people often spend 80% of time outside the office visiting customers and developing account strategies with salespeople. Careful selection of customer-marketing people is important. Marketing managers must choose those people who can remain positive and optimistic amid a hectic lifestyle. Customer-marketing people must be able to juggle a lot of customer issues simultaneously. People who like to deal with one problem at a time are not going to be successful in customer-marketing positions.

ROLE OF TECHNICAL MARKETING

There are marketers who love the engineering aspects of their work and can't leave their technical side behind them after moving into a marketing role. In some cases, there are also former engineers who have moved to customer- or product-marketing roles and have determined that they don't enjoy the business side of their jobs. Dealing with customers is not for everyone. People who favor the technical side of marketing can find good homes in technical marketing roles. The technical marketing people are primarily responsible for defining new products and writing marketing requirement documents that can be used by product-marketing people to bring new products to market. The technical-marketing people are often highly respected by engineering teams, as they have a better understanding of what engineers need to do to get their jobs done.

Technical marketing people do not have any revenue responsibilities or any ongoing customer management activities. On the other hand, they are measured by their ability to analyze competitors' product lines and communicate to marketing organizations how they can bring new and innovative products to markets that could beat the competitors' product offering. Technical marketing people are also responsible for developing the preliminary product guides and manuals used by application groups to develop final product user guides and datasheets.

In some case, overlaps exist between product marketing and technical marketing roles. Product marketing staff may want to be more involved with early product development work and may see the technical marketing people's role as a duplication of effort. It's also an issue with both product- and customer-marketing people that technical-marketing people have no real revenue or design-win targets. In some sense, technical marketing people are viewed as people who have the perfect job, one without the daily pressure of delivering quarterly or yearly revenue targets.

NEED FOR MARKETING LOGISTICS

In some companies, in addition to product marketing people, there are also marketers who spend most of their time managing the supply and demand of product lines. These people often do not have an engineering degree, but they have the crucial role of monitoring customer orders and revenue and comparing these numbers to forecasts. Marketing logistics people spend most of their time making sure that customer marketing, product marketing, and salespeople agree on revenue targets, and gauging realistic forecasts for each quarter. Marketing logistics people also work closely with operations people to make sure products are shipped on time and that salespeople have a contact whom they can connect with to find out the status of product shipments and inventory.

Marketing logistics people also maintain product lines' forecasts and continuously work to make sure adequate wafers and packages are ordered for timely product shipments to customers. It's important to know that, without a dedicated marketing logistics group, customer-marketing people or, in some cases, product-marketing people may have to spend a considerable amount of their time dealing with operations people making sure that customers get their parts on time.

ROLE OF APPLICATION ENGINEERING GROUPS

One of the key groups within a marketing organization is the application engineering group, which is tasked with providing pre- and post-sales customer support. An application engineering group's ability to service customers can make all the difference in how customers view a company. The attitude and responsiveness of application groups is what customers remember and measure companies by. Having the right people working in an application engineering group is important — a marketing manager needs application engineers who are service-minded and do not mind getting interrupted throughout their workday to help salespeople or the end-customers with their problems.

People working in application engineering groups must be highly technical and must have the ability to learn a product's features and functionality quickly. In some cases, many application engineers obtain their business degree so they can eventually move to either a technical-marketing role or other marketing roles. A marketing manager must ensure that application engineers stay current with the latest product features and provide complete support for customers.

Application engineering groups are also responsible for maintaining all product collaterals, such as data sheets, reference designs, and applications notes. In some cases, they also develop hardware and software to showcase product features and performance. In addition, they perform competitive analyses on competitors' products. Performing competitive analyses on competitors' products gives application engineers insight on how customers perceive competitors' products. Managers must encourage application engineers to know that they are part of the marketing organization and that they can't ignore the competitive nature of the business.

The other responsibility of an application engineering group is to provide customer and salespeople with product training. Normally, application engineers are based both in the field and in the factory. Application engineers in the field are often part of the sales group, while those in the factory usually work in marketing organizations. Either way, the factory-based application engineers communicate closely with their field-based colleagues and directly with customers. In general, field-based application engineers rely a great deal on the factory for information and support since they are on the front line of customer support. It's up to the factory-based application engineers to make sure their counterparts in the field are getting the necessary information, support, and training.

A marketing manager must ensure that his application engineers are strong in order to guarantee excellent sales and customer support. Poor customer support is the primary cause of customers' and sales people's negative perception of a product line. Sales people will lose confidence in a product line and give up trying to promote the products in if they lose confidence in the products' technical support.

Marketing managers must ensure that application engineers have the right skills and training to handle a product line. For example, if a product has a large software component, application engineers must be familiar with the development and support of that software. Weak application engineering is a detriment to winning customers, which could also impact the work of development engineers. Weak or under-trained application engineering groups will rely on development and product engineers to support customers. While in some cases this cannot be avoided, constantly depending on development engineers (whether they are software or hardware) for customer support will damage a company's ability to execute on launching new products.

HIRING THE RIGHT MARKETING PEOPLE

Hiring new people in the semiconductor industry is always challenging. However, it's even more difficult to find the right people who have both business and engineering skills to successfully market semiconductor products. Many semiconductor marketers have a tendency to gravitate toward their technical side, even after moving to a marketing role. They often try to engineer their way out of business challenges and fail to spend enough time developing the interpersonal skills needed to be successful in the business world.

To hire the right people for a marketing group, a marketing manager must first determine what it is that he wants a new hire to do. As was previously noted, a manger must first organize a marketing group based on the product categories and customer segments he plans to serve. After he has established an organization, he must make sure that existing employees are doing the work for which they are best suited. He can then fill the gaps by hiring new people from outside the company, or in some cases from within the company.

SUMMARY

Marketing managers must invest the time needed to hire the right people in order to form effective marketing organizations. The managers, in some cases, should act in the same fashion as military commanders. In a market there is usually an enemy as well as battlefield objectives. Only the right people organized effectively will yield positive results. A manager, just like a commander, will take the time to evaluate his people and decide how their skills are best utilized to benefit the entire team's effort. While he may try to accommodate people as he can, ultimately he cannot compromise team effectiveness in the process. Once he has organized his team, he must communicate the individual responsibilities of team members and make them well understood to achieve the objectives of the team

getting help ›

Semiconductor marketing is not for the faint of heart. Often, one must seek help or guidance to get through tough business problems. There will come a time when a marketer may need to get help from others to help make decisions about issues such as revenue generation tactics, product or business strategy, and organizational problems.

A part of a marketer's success depends on his ability to seek and find advice from people around him. If a product line is in trouble, or if there is a difficult business problem to solve, there is no shame in seeking help. It may be easier to sit back and hope that problems will work themselves out over time. But, hope is not a strategy.

This is especially true for people new in the marketing field. Young marketers often do not realize how much they don't know. They often lack the necessary experience to rely on their intuition. A marketer must establish and nurture a network of close business colleagues, both inside and outside the workplace. Younger marketers are especially in need of mentors and colleagues who can help them. These mentors should include people who have spent many years in the industry and have weathered industry downturns as well as boom times.

SEEKING HELP INTERNALLY

The first place one should look for help is inside his company. Developing close relationships with business colleagues is a necessary part of having a successful career. When faced with product line related issues, a marketer should first approach a few people who have similar or greater responsibility and get their views and opinions.

Company colleagues can often be a source for a wealth of knowledge. Not only can they provide good advice about marketing and business, but they may also be plugged into the company leadership. Contacts like these can help a marketer determine how management views his product line. It's beneficial to know the "lay of the land" before taking action.

Due to the dynamic nature of the semiconductor industry, companies are forced to continuously reorganize their product lines and the make-up of their marketing groups. This means that management can discontinue or realign the focus of a product line. A marketer's management and promotion of a product line must also include the development of key relationships in the company to ensure the buy-in of management on the strategic direction of his product line.

Some larger semiconductor companies have departments that address product-line strategy at the macro level. These groups are often called "business development" or "strategic planning" groups. The CEOs and vice presidents of companies tend to work closely with business development groups because they often have extensive experience in the industry. Marketers should also establish relationships with the key people in the business development groups. Marketers should consult the business development people about the prospect of their market segments, product lines, the importance of the product lines to the company, and their performance as the head of the product lines.

A marketer should also use "offsite meetings" to harvest the views of a broad group of key engineers and managers about his products. "Offsite meetings" can be used to discuss strategy or the current status of business and its potential. These meetings are a great way to get the views of influential people regarding the current health and future of various product lines. Marketers should make certain that the offsite agenda is clear and that they are prepared to articulate a clear view of their product line. The offsite meetings are also a great forum for high-level marketing and engineering people to get to know each other better, and realize they all work for the same team. Many companies use offsite meetings to devise short-term fixes to long-term problems. An offsite meeting should always be followed with decisive actions. Otherwise, a marketer runs the risk of being viewed as weak and indecisive.

Sometimes the best way to get an unbiased view from people who can be trusted is to seek the advice of your friends in the industry, industry consultants, or consulting firms.

GETTING HELP EXTERNALLY

When faced with a challenging strategic issue, a marketer (or more likely a marketing manager) should consider seeking outside help. Unofficially, outside help could come from industry friends and former colleagues. People who work in marketing, sales, and management in the semiconductor industry have to stay on top of industry trends and are constantly exposed to macro- and micro-level events that can shape the future of their products or industry. By keeping in touch with friends in the industry, a marketer can obtain unbiased views about strategy and business problems.

A marketer should seek the advice of friends who are acting, or have acted, as general managers or vice presidents in similar industries, because they are experienced in dealing with more strategic issues than day-to-day tactical issues. In some cases, these discussions could even lead to new business deals. The deals could involve selling a product or a product line or even buying a new product line, allowing a marketer to fill a hole in his product portfolio. A side benefit of getting help from friends is that it does not cost a lot.

One problem though, with consulting with industry friends is that it is hard to be completely open with them without divulging company confidentialities. A marketer can solve this problem by officially hiring an industry consultant to serve as a consultant via a formal contract. A contract can include a non-disclosure agreement. It also serves to validate a consultant's opinion with other members of a company. This will not work, of course, if the consultant is also doing work for the competition.

Besides individual consultants, there are also small consulting firms that marketers could tap into to get help. Smaller consulting companies can consist of a handful of people who have worked in a specific semiconductor industry segment and may even be providing research data and news to supplement their consulting work. These consulting firms have a tendency to stay very close to all the industry dynamics and issues affecting a product line and can help with focused consulting jobs that do not include massive reorganization or strategic acquisitions or divestitures. The consultants of small firms may have good insight with competitors and could provide some insight into the public information available on competitors' plans.

A massive reorganization or major strategic realignment possibly requires the help of a larger consulting company. In fact, the decision to hire a large consultancy is usually made by the president or CEO of a company. Large consulting firms can offer a wide array of resources needed to analyze the current state of a business, provide recommendations about the direction of product lines, as well as provide advice on company-wide re-organization.

The consultants from large firms, in most cases, have to learn the specific markets and products a company covers. Individual marketers need to take this process seriously or otherwise run the risk of being on the wrong side of a reorganization.

Because consultants from large firms are hired and report directly to a CEO or president, their views are highly valued by them. A marketer should give the consultants full support and help them to come up with an unbiased view about the dynamics of industry segments and his product line.

Consultants get their knowledge by talking to various groups inside the company and by talking to the company's customers. They also have access to a large amount of research reports done by others in their firms, or by independent smaller firms and individuals.

If, at the conclusion of the consultants' work, the company decides to reduce future investment in a product line or even divest a product line, good marketers should be supportive of the decision. However, if a marketer truly believes in the future of his product line and has data to support his stance, he should go out of his way to save the product line.

In some cases, no matter how much a marketer believes in his product line, senior management may have already decided to shut it down for financial or even political reasons. Some of the reasons may even be confidential. In these cases, the marketer is better off supporting the decision and helping in the reorganization or the divesture of the product line.

MARKETING TRAINING PROGRAMS

Unlike most engineers working in the semiconductor industry, marketers working in the industry do not get a chance to continuously upgrade their marketing skills. Most semiconductor marketers don't think they need any training. They think that just because they know all about their product line and the markets they serve, they don't need to know about high-level industry dynamics and various marketing trends that could help them better perform their job.

As is the case in every other profession, semiconductor marketers should take various business-training classes to get fresh views about marketing and semiconductor industry trends. Semiconductor- marketers should earn an MBA, if they don't already have one, in order to develop a broader understanding of business issues and challenges facing their product line. However, even after getting their MBA, it's important for marketers of semiconductor products to stay fresh and on top of trends shaping the future of the industry.

Besides pure business training courses, it's also important for semiconductor marketers to attend technology seminars that mix pure engineering training with an emphasis on business aspects of new technologies. This allows marketers to make the long-term decisions needed to grow a semiconductor product line. Marketers must also take the time to understand the latest in semiconductor process technology, EDA technologies, chip design techniques, and software design issues. This knowledge can be very useful, especially in the development of complex SoC product road maps.

One way to provide systematic training for those interested in semiconductor marketing is to bring the training to the company. There are technology marketing consultants who can give marketers and engineers fresh views on the latest marketing techniques and trends. These internal training programs also provide marketers and engineers with a forum so they can share their thoughts. Their interaction in a training setting allows them to develop a common understanding of issues and will lead them to work better as a team.

If a company cannot afford to bring lecturers in from the outside, there are also many training programs and seminars available outside the company. A local university can often provide general marketing or business related courses. The challenge is to find industry-specific training programs that provide up- to-date and applicable information about industry trends and dynamics. Unfortunately, business schools do not have dedicated marketing classes that address the training needs of semiconductor marketers.

To get the latest semiconductor industry trends, it is important to read various industry research reports prepared by both small and large firms. Many companies do not see value in paying for such industry and product-specific research reports. This is a mistake. These reports can help a marketer understand the markets he serves. In addition, many customers may formulate their thoughts about a product or a company based on the content of industry research reports. By reading and understanding these reports, a marketer can organize promotional efforts to get favorable placement in them.

A marketer should read extensively about the business world in general. Just as with a product, a marketer has to know his strengths and weaknesses and continuously work toward matching his strengths with available opportunities.

It's our hope that, after reading this book, readers understand that semiconductor marketing is a complex discipline.

BECOMING A ROAD WARRIOR, HELP ON TRAVEL

Traveling is something most marketers do a lot. In some cases, an active customer marketing person might be on the road every week. Considering that marketers can spend so much time on the road, it would be negligent not to provide some travel guidelines to readers on this subject.

As previously discussed, meeting with customers and salespeople in the field is key to a marketer's success and a critical component to the development of product strategy. Marketers travel more than any other group in the company, except senior sales VPs. But unlike sales VPs, traveling for a marketer could also be a detriment to relationship building and product management that require daily guidance in the factory, especially for product line managers working in a vertical organization. If a marketer is away from the office too much, then product development efforts can suffer; on the other hand, never traveling is a detriment to the selling efforts of products. Traveling to customers and regional sales offices also offers a tremendous opportunity to hear firsthand feedback about products and road maps. Finding the right balance on travel is very important.

To optimize the time spent on any trip, marketers must always have clear objectives. It is amazing how many trips typical marketers take without taking the time to develop a list of objectives for the trip. A marketer must make certain that not only he, but also the salespeople understand the objectives of a trip and have bought into the plan. After getting the buy-in of the salespeople, a marketer can ask them to set up a series of meetings to visit the customers in their region.

Travel time away from the office has to be managed so that engineers don't have to wait for a marketer's return to make decisions and move forward with product development objectives. Marketers must be reachable via phone and email throughout their trip.

The time away from the office also means time away from family. Traveling extensively can have negative effects on family relationships. A marketer may need the support of extended family members during long haul trips to help his immediate family manage their lives. Setting up trips in advance will help reduce the stress of leaving the family by allowing them to prepare. The habit of traveling within a moment's notice is a sign that a marketer is putting work ahead of family, a practice that could have disastrous consequences. In our view, a job is never worth the value of a family.

When flying, never check luggage by packing the least amount possible. Checking baggage can slow a traveler down and/or cause missed flights. Not checking bags allow a marketer to make "on the road" changes to an itinerary without worrying whether baggage will eventually catch up.

Flying can provide a marketer with the rare opportunity of being away from the phone and email access. It is a great time for reading and reflection. A marketer can use this time to think about a product lines' high-level issues, as well as work on revenue targets and next year's product goals and objectives.

SUMMARY

Successful marketers seek guidance to help in critical decision-making. As a marketer develops in his career he will find that his contacts, mentors, and consultants will provide an ever-growing advice network. Successful marketers make it a point to develop this support network.

Marketers also need to seek out training opportunities. The material learned in training makes a marketer more effective in his job. The training as well as the time away from work contemplating business and product problems can help yield new strategies and solutions.

the successful marketer ›

Finding good marketers in the semiconductor industry traditionally has been very difficult. It is not by accident that semiconductor marketing is one of the highest-paid positions in the technology sector. Semiconductor marketing roles are hard to fill because the personalities of people who want to work in the semiconductor field are generally not well suited for marketing. After all, the one prerequisite for the semiconductor field is an engineering degree. Most people who graduate with engineering degrees want to be engineers. Others who are extremely extroverted usually end up in sales. So who is left for marketing? What attributes and skills does a person need to succeed in marketing?

IMPORTANT ATTRIBUTES

There is more to good marketing than knowledge of the profession. Marketing requires the practitioner to have or develop certain attributes. In addition to knowledge, good marketing relies on intuition, charisma, interpersonal skills, and experience.

Most semiconductor marketers begin their careers in engineering and later move into marketing roles. For some, the transition to marketing seems natural. For others, it is very challenging because the fundamental approach to problem solving in engineering is different from what is required for marketing.

In engineering, one is constantly dealing with the science of deductive reasoning. Electrical engineering, like most other engineering fields, is about designing elements and components that perform a set of deterministic functions or tasks. As was discussed in previous chapters, marketing also attempts through segmentation and evaluation to make a more deterministic analysis of markets and products best suited for those markets. However, there is a significant difference between engineering analysis and marketing analysis. Since markets are a composite of human behavior, they tend to be chaotic and complex.

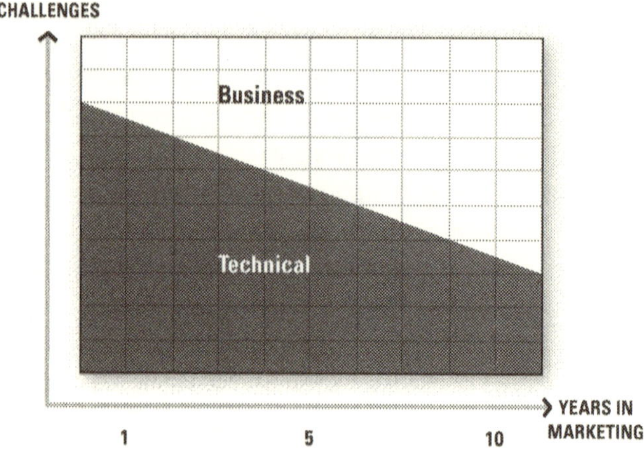

For this reason, marketing relies heavily on inductive reasoning. Marketing problems are not always deterministic. Courses of action are not easily predicted or compared. Due to the "gray" nature of the field (as opposed to the "black and white" nature associated with engineering), marketers have to rely on a host of new skills to help fill in the blanks.

One of these skills is interpersonal ability. In contrast, engineers do not have to develop their interpersonal skills to be able to have a successful career in engineering. But marketing requires good people skills. Marketers have to sell their ideas to customers as well as people inside their own company. Marketers also need relationships to gain access, collect information, learn new ideas, and influence events and decisions.

Creativity is also required for successful marketing. Creating is the process of making something out of nothing, and it is a quality that a marketer has to tap into to help overcome strategic and tactical challenges. Strategically, creativity is used to come up with profitable product and business ideas. Tactically, a marketer needs creativity to solve a constant stream of business problems and negotiations that are part of the job.

In marketing, creativity has to be coupled with leadership skills. A marketer must be able to persuade and convince management and engineers to accept his ideas and, in many cases, to support the ideas and work hard to fulfill them. Whether he is advocating a product idea, or suggesting an approach in a price negotiation, a marketer will constantly draw on his ability to create solutions and carry out plans through leadership.

Another characteristic of a marketer is his general curiosity about the world and about people. Creativity and curiosity go hand in hand. A marketer needs to take in ideas around him in order to generate new ideas. Marketers feed their curiosity through reading and through conversations with a wide variety of people.

At a minimum, good semiconductor marketers read about business and technology trends. Learning about how businesses are formed and what makes businesses successful is required for success in marketing. Understanding technology trends and considering their implications are also fundamental.

MOVING FROM ENGINEERING TO MARKETING

Most semiconductor marketers start their careers in engineering, and at some point they switch into the marketing field. Many marketers realize their career paths early on and switch into marketing after only a few years as engineers. Others are promoted into engineering management to the point where they take on business responsibilities and eventually decide to move to marketing. So how do you know if you are destined for the business side of the industry?

The first required element is a desire to switch to the business side. This desire is usually fueled by a combination of interest in business and frustration with pure engineering work. An engineer's frustration often stems from being moved from one project to another, or from having little or no understanding of the market value of his work. Many engineers want to find out for themselves what drives technology investment decisions.

For those engineers interested in moving into marketing, the pathway is not always clear. But each type of engineering discipline, whether it is product, design, or systems, uses skills that could be useful in the marketing field. However, there are also engineering habits that could be detrimental in marketing.

FROM CHIP DESIGN TO MARKETING

For example, let's assume a chip designer is interested in moving into marketing. Clearly, this background is of value in the marketing field: a chip designer has firsthand knowledge of the design process and a complete understanding of what it takes to develop ICs. This is valuable in marketing, where one needs to work closely with engineering management to understand resource and schedule impacts on programs. Also, having expertise in a specific technology is useful when presenting to customers. But, an experience in chip design generally does not expose one to customers, product marketing, or business issues. In addition, chip designers in general only work on very small components of an overall system. Chip designers may not appreciate the entire system solution or what drives that solution, such as standards and customer demands.

Chip designers might consider a move first to systems engineering before proceeding to marketing. A systems engineer has to be on top of trends and competitors in the marketplace, thereby developing product management skills. In addition, systems engineers are often brought into strategic planning sessions as experts to discuss issues directly with marketing and customers. After a few years in systems engineering, a natural move is to take a technical marketing position to increase exposure to business issues, standards bodies, and customers. From there, a budding marketer can expand his business acumen and eventually move into product or strategic marketing.

FROM PRODUCT ENGINEERING TO MARKETING

Another group of engineers with useful marketing skills are product engineers. A product engineer works with all other engineering groups to ensure a chip is designed properly in order to ensure low costs and quality of manufacture. That means product engineers often end up working side by side with semiconductor marketers through the life of products. They often observe firsthand what marketing people do. In addition, product engineers are involved with marketing to drive product cost-reduction plans. In a switch to marketing, product engineers bring broad and invaluable experience and knowledge about manufacturing, yield, test, and the semiconductor supply chain process.

Also, product engineers learn how to deal with large cross-functional teams during the development of a product — an important skill in product marketing as well. In addition, product engineers get a lot of customer interface, often helping marketing to communicate product-yield and quality issues to customers.

FROM APPLICATIONS ENGINEERING TO MARKETING

Another group well poised for marketing is application engineering. In some instances, application engineers are already part of sales and marketing groups. These engineers provide the connection between customers' technical groups and internal engineering groups. In addition, application groups are often tasked with attending industry shows and helping marketing in demonstrating products, or helping with presentations to customers (e.g., discussing product features). Clearly, the combination of customer exposure, technical knowledge, and a close working relationship with marketing provides application engineers with a great background for moving into marketing.

There is another reason why application engineers move into marketing. The careers of many application engineers are often capped at a certain level. Salaries are high for application engineers, but growth is limited. There are not many managerial jobs available to application engineers in the industry. It is not uncommon for an application engineer to earn an MBA and then move into marketing. Application engineers have the benefit of seeing all aspects of a marketing job before deciding if that is what they really want to do.

MOVING FROM SALES TO MARKETING

Finally, it is not unheard of for salespeople to move into marketing. Of all the people in a semiconductor company, salespeople spend the most amount of time with marketing groups. Salespeople work closely with various marketing groups to service their customers effectively. It is common for salespeople to switch from sales to marketing (and vice-versa). In some companies, sales and marketing groups report to the same vice president.

However, big differences exist between sales and marketing. Salespeople are in the field trying to close business. The sales role is very tactical and requires people with strong personalities to persuade and convince customers to hand over their hard-earned money in exchange for a product. A move into marketing from sales allows a salesperson to concentrate on understanding a product line and how it fits into the bigger picture. The position also allows a salesperson to think a few years out, instead of just one or two fiscal quarters.

Salespeople have some natural advantages, including interpersonal skills and customer interaction. Another advantage is that salespeople should have no trouble working with other salespeople once they make the switch to the marketing role. The areas of challenge for a salesperson will be learning the program management side of marketing. Salespeople need to learn team building and must work with engineering groups as well as internal management to sell their product ideas. Usually these efforts involve negotiations and compromises in terms of resource allocations and schedules. These aspects of the job could be difficult for salespeople who are more action-oriented and may have difficulty adjusting to the tempo of a complex engineering organization.

Another challenge for salespeople is dealing with the formal process of long-term product planning. Salespeople often are focused on short-term results and may not appreciate the importance of thinking many years down the road. Sales is all about selling what the company has today, turning current available products to revenue. Nevertheless, salespeople who are interested in making the transition are often well rewarded. Generally, they have what it takes to be successful in the marketing field and, in addition, are well positioned to move up in a company with their combination of marketing and sales talents.

CONSIDERING A MOVE TO MARKETING

Simply stated, promotion and career growth are good reasons to consider a move to marketing. People interested in running companies would benefit from having five to ten years of marketing experience. By definition, those who start or run companies need to spend time thinking about product plans and product road maps. The concept of turning ideas into money is fundamental to the activity of marketing. So, those who have any desire to become a CEO or to start a company should spend a few years in marketing to gain some useful experience.

Often, when we are faced with a career move, we ask ourselves if we have the right stuff to be successful at that career. Some people take tests to determine what career matches their personality better. But, those tests usually are generic tests and do not address semiconductor marketing specifically; they are designed to provide a basic idea whether a person has a tendency to gravitate toward science, marketing, or art.

A good way for someone to determine if he is truly poised to make a move to marketing is to observe his own behavior and interests. For example, does the person have an interest in business issues? Does he read business magazines or have an interest in business documentaries on TV? Is he happy in his present position or frustrated? Does he enjoy the company of business people? Is he fascinated by what drives businesses (e.g., such as entrepreneurial ideas, product planning, and investment strategies)? Does he consider himself an engineer for life — or he sees engineering as simply a stepping-stone to something else. In some cases, the answers to these questions will point to obvious directions. If they do not, one should be more circumspect about changing a career.

People should realize that moving from engineering into marketing constitutes a dramatic career change. If nothing else, a move into marketing will require an investment in clothing. The ten-year-old green suit from a university graduation will have to be upgraded to more professional threads. For those people who spent too many years in a dark room in front of a Unix machine, it is time to consider a shave and a haircut — and of course a move to using PCs.

Perhaps the hardest thing for engineers to give up when they move to marketing is their technical expertise. Once engineers move away from engineering, they will start to lose their engineering expertise because marketing demands a much broader view. Engineers have trouble with this because they were once identified and got emotional support for being an expert in a particular technical area.

Those engineers, who decide to move to the business side, must let go of their attachment to their engineering past. Holding on to the old glory days is detrimental to finding success in a new business role. One can always return to engineering and redevelop a technical expertise if marketing does not work out. But to make it work out a new marketer must devote himself to the profession.

Another way for people to consider if semiconductor marketing is the right career choice is to determine if engineering was their choice to begin with. Parents often influence their children to study engineering because of the stable employment opportunities provided by engineering work. This is especially true of foreign-born engineers. This may answer why there are so many foreign-born people in the semiconductor field. In developing nations, engineering is considered a top profession. So, it is natural for parents to promote engineering as a career goal for their children. A good starting point in a self-evaluation is to consider the reasons why one went into engineering in the first place.

NECESSARY BACKGROUND FOR MARKETING

As was stated at the beginning of this chapter, finding good semiconductor marketing people is difficult. The reason is that it requires people who are technically astute combined with leadership and business skills. Usually it is easy to find technical people and business people — but the combination is a challenge.

For those interested in meeting this challenge from outside the field, a technical degree is a prerequisite. There are exceptions to this rule, but for the most part it is preferable to have an electrical engineering degree or at least some kind of engineering, computer science, mathematics, or physics degree. Since semiconductor marketing requires daily contact with technical issues, technical people, and technical problems, having an engineering or a technical degree is necessary in order to communicate and make decisions.

Even if a non-degreed person has been given a product marketing responsibility for a semiconductor product line, that person must seriously consider getting an engineering degree or seek employment with a larger company such as Intel. Intel and other large semiconductor companies have large branding and promotional activities that may not require all marketers to have technical backgrounds.

In addition to having an engineering degree, an MBA is advantageous. Even if it takes many years part-time to earn the degree, graduating from a formal MBA program will help one excel in semiconductor marketing. Without an MBA, one may face difficulties having effective business discussions with higher-level executives. Also, a formal MBA program provides a great opportunity to read and learn about factors that influence not only the high-tech markets but also the non-high-tech markets.

Besides having the necessary technical and business education, semiconductor marketers must also continually maintain their knowledge of business world through reading general (as well as high-tech) journals and books. Reading biographies of successful business people, management books, financial magazines, and newspapers should become a daily activity.

SUMMARY

There are two schools of thought regarding the semiconductor marketing discipline. There are those who argue that marketing is generic and can be applied to any industry. This group believes that if someone marketed soft drinks, he can also market semiconductor products. This group also argues that an engineering background is not necessary for semiconductor marketing and may even be a detriment. The authors do not agree with this school of thought.

We believe that semiconductor marketing requires a thorough knowledge of the field from the technical, operational, as well as business perspectives. It is true that there are many marketing people at Intel with no engineering background who do a fine job at Intel product branding. However, most semiconductor companies do not have or need a comprehensive branding strategy like Intel. This means that there are very few opportunities available in semiconductor marketing for those who do not have engineering backgrounds. Semiconductor marketing requires a person to know as much as possible about the semiconductor industry and the technology behind various product categories. However, the technical knowledge must be accompanied with the softer skills.

CHAPTER 16
advice to new marketers ›

One of the goals of this book is to serve as a marketing primer to those outside of the marketing field as well as to those new to semiconductor marketing. This chapter is dedicated to those marketers who have had less than five years experience in the field and provides advice on how to be successful. This chapter is useful for non-marketers as well since it discusses the approach successful marketers use to solve problems. It also discusses habits of success that are useful in any profession.

Successful marketing depends on a marketer's leadership skills as well as his ability to analyze markets, execute on the right product strategy, and serve as an effective business manager.

illustration 33: marketing activities

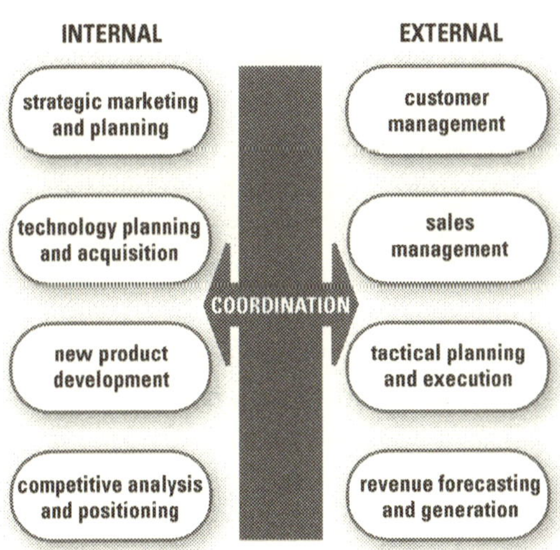

Success ultimately depends on how an individual can harness relevant knowledge, develop leadership skills and execute on a plan. A leader must by definition convince others and be able to sell his ideas. This chapter examines how a marketer new to the field can help ensure his growth and success in the marketing field.

DEVELOPING EFFECTIVE MARKETING HABITS

It is important to make goals. But true success comes from making a habit out of an activity that guarantees positive results. For example, if one's goal is to be trim and fit, it is not enough to say, "I will go to the gym three times a week." While one may have enough self-discipline to do it for a few weeks, these types of efforts usually fail. It is easy to get tired or bored with the routine. People can become despondent that results do not come about as quickly as expected.

It is easier to reach a goal when a success-oriented routine becomes a habit. This was the main message of the book Seven Habits of Highly Effective People by Steven Covey. Note the term "habits" in the title. Being fit is a great goal. But to make it a reality, one needs to make exercise and healthy eating a habit. Rather than concentrating on the end result, it is better to concentrate on the habit. The habit becomes the goal. One should consider exercise like any other habit — its just something you do, possibly on a daily basis if appropriate. Like a smoker makes time for a cigarette (bad habit), a healthy person makes time to exercise (good habit).

The effectiveness of habits is that once they are ingrained in the psyche, they become a powerful force driving one to an end. Success oriented habits practically force people to be successful over time. In addition, by focusing on the habit, a person will be less inclined to obsess about reaching a goal in the future. The habit becomes the fulfillment of the goal and the future takes care of itself. There is no disappointment with temporary failures along the way.

An effective marketer makes a habit of success-oriented activities that expand knowledge, improve decision-making and business skills, improve speaking and interpersonal skills, and sharpen leadership skills. Successful marketers make a habit of learning. Reading as much as one can about business, entrepreneurship, management, technology and finance is a must. Taking advantage of public speaking to improve skills and learn how to formulate thoughts in a cohesive manner and express them to others is fundamental.

Through habits, it is useful to attack those areas that are important but where one's skills are weak. For example, a marketer who is weak in forecasting and tracking of business should make a point of learning this skill. The worst thing he can do is to avoid the forecasting process, especially if it is a necessary part of the job. Instead, he should dive into it. He should spend a few minutes every day to update his forecasts. He should pride himself on having the company's most comprehensive forecast. He should talk to people who are knowledgeable in forecasting and seek their advice. He should think about ways to improve sales communications and forecasting methodology. If he makes forecasting a habit, it is just a matter of time before he masters the skill.

Forecasting is just one example. If a marketer has a weak background in negotiating, he must take special interest in obtaining effective negotiating skills. In general most people hate to negotiate. However, it is a critical skill in business that marketers must inculcate negotiating into their daily habits.

It is easy to find opportunities to bargain every day. Try to pay the least amount of money for a product or a service and sell it at the highest price possible. Swap meets and garage sales are a great place to bargain either as a seller or as a buyer. One can bargain at hotels, or when buying a car or a house. One can try to bargain in situations where no one usually does, like at a convenience store or a bank. It's often surprising how many times people are willing to offer a discount. During business travels one should always try to negotiate to get free upgrades on cars, hotel rooms or on airplanes. When asking for discounts and upgrades, there is no downside. The worst one can get is a "no;" and end up where one started. It does not cost anything to ask. The same is true when a marketer is trying to sell a semiconductor IC or negotiating a contract with a customer. A good marketer will always try to find a way to get more, and at the same time he will strive to be fair by understanding what the other party needs or wants.

ADOPTING A MARKETING MIND-SET

In general, a lot of good marketing is based on good intuition. Good intuition will help guide a marketer's actions and let him know when there is trouble, or when he is on the right track. Intuition stems from experience and understanding. Intuition, coupled with creativity, curiosity, people skills, and leadership are the basic attributes of a successful semiconductor marketer.

The development of skills and habits for success is crucial to the goal of effective marketing. But there are also "habits of thought" that set marketing people apart from those in almost every other function in the company — a marketing mind-set.

What is this mind-set? It is marketing theory expressed through leadership, communication, and business decision-making. The mind-set is a way to approach problem solving and decision-making that is grounded in business theory and expressed or executed through leadership skills.

SOME EXAMPLES OF MARKETING MIND-SET INCLUDE:

Thinking in terms of "return on investment" (ROI). A marketer should never consider an opportunity, a solution to a problem, or a product investment without considering the ROI. To a seasoned marketer, the ROI should be a knee-jerk reaction to almost every business issue that comes up. This of course should not be surprising. After all, business decisions for the most part are concerned with the notion of making money. Making money means that the value of your return is greater than the value of your investment. Marketers should think in terms of ROI, regardless whether it's done in their head, on paper, or on a complex spreadsheet.

Creative thinking. When faced with problems the effective marketer should consider multiple options. The term "thinking out of the box" is a bit overused, but it is descriptive of the concept that solutions to business problems may exist outside the normal parameters of one's thought process. The mind-set also is about constantly combining technology considerations with business opportunities or demands. A vast amount of information must be processed quickly, requiring one to "connect the dots" to come up with not-so-obvious conclusions while combining business and technical knowledge.

Optimism. Looking at the bright side of life is another important element in the marketing mind-set. A marketer who has done his homework, has understood his markets, and has worked hard to define products should by definition be optimistic. Optimism is also a necessary ingredient in selling and leadership. A company will not follow a marketer's ideas if he is pessimistic and questioning his own decisions at every turn. When a marketer has worked hard to define a strategy, he must have faith that he is on a path to glory. Optimism does not mean reckless confidence or blind hopefulness. A marketer must still observe the market and watch for signs of change.

Paranoia and skepticism. While optimism is a key to the engine of marketing success, marketers must also recognize that a lot of forces are working against success. It is possible to be a little paranoid and skeptical while being optimistic. Marketing people must develop an intuition about the truth of the information they receive. To put it another way, marketing people must be good malarkey detectors. Marketing requires the ability to take in a lot of information fast and determine the validity of what one reads and hears. There is a huge amount of disinformation out there.

Being a good malarkey detector also means one questions others in their follow- through. People tend to fill in their gaps — whether it is gaps in information or gaps in execution — with malarkey. Marketing people develop an intuition in their mind-set that roots out this malarkey in their quest for the truth. Good marketing people also try to limit the malarkey they themselves generate. Marketing is rife with opportunities to generate malarkey. A good marketer says, "I don't know," when he or she really does not know, rather than respond to a question with incomplete information.

Multitasking. The capacity to multitask is also a key part of the marketing mind-set. Engineers moving into marketing must adopt multitasking to be effective in marketing. Engineers solve problems by breaking them down to manageable pieces and solving them in a serial fashion. While this is an effective means of problem solving (even for marketing problems), there is rarely the luxury to concentrate on one task to completion for the marketer. The world cannot wait for a marketer to work in a serial manner. Marketing demands multitasking of activities and requires working on projects that sometimes have very gray milestones.

UNDERSTANDING PERSONAL SUCCESS FACTORS

Career growth in semiconductor marketing primarily depends on how effectively one can grow a business or maintain a business in a shrinking market segment. It also doesn't hurt to be at the right place at the right time. Unlike engineering, where design work speaks for one's ability to innovate, in marketing one is judged by the ability to present and articulate thoughts, analyze abstract market dynamics, and have the personality that makes people trust one's judgment. Therefore, all of a marketer's actions can create either a positive or a negative perception in the mind of those who can strongly influence his career path.

A marketing job often feels like an audition where one is constantly presenting in front of invisible judges. Career growth depends on a marketer's ability to persuade others that he can be a business leader and be given higher levels of responsibilities. Most high-tech CEOs come from the marketing side of the business. This means that a career in marketing could provide one with the opportunity to someday run an entire company. However, before getting there, a marketer needs to set his goals high and develop the skills needed to create the right career opportunities. A marketer can't depend on others to give him what he deserves. As part of the work, a marketer must spend time marketing himself and building trust among colleagues and managers on his ability to lead a product line or a business unit.

It is also wise for marketers to seek direct and indirect feedback on their performance and what can be done to improve chances for getting higher levels of responsibilities. However, at some point, a marketer may have to make the choice to find opportunities in a different division within a company, or even with a new company, if his career has stagnated. A marketer cannot remain with a company that does not appreciate his abilities or refuses to provide the opportunities that can allow him to showcase his skills.

A marketer who decides to leave a company should do so without burning any bridges in the process. There are always jobs out there for good people. If a marketer who is dissatisfied with the job decides to stay on anyway, then it is best to concentrate on the work and avoid too much griping. People do not like to hear complaints. Ranting and complaining will limit career growth as it stains an image and reputation. The semiconductor community is a small world and a bad reputation can get around very easily.

illustration 34: semiconductor marketer's career roadmap

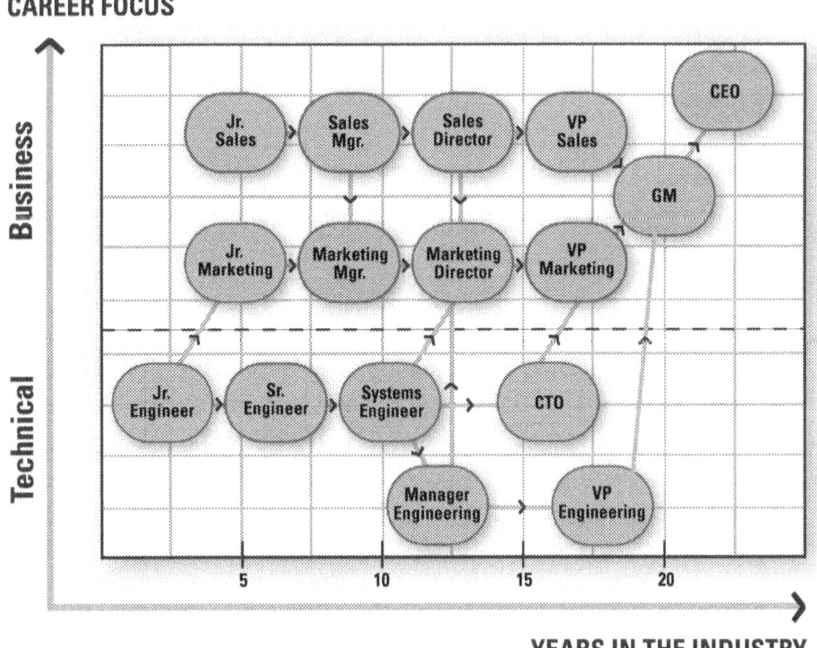

ADVICE TO NEW MARKETERS >

To grow in marketing, a marketer in charge of a single product should set his sights on becoming a product-line manager responsible for a family of products. Depending on the growth of the business, one must then look for a way to become the director of marketing, responsible for all aspects of running multiple product lines. Depending on the size of a company, the opportunities to grow beyond director may be very limited. A director-level marketer may have to look outside his company to further grow his career.

When considering a move outside the company, one should consider start-ups. Start-ups can offer a fresh chance to showcase marketing skills. At a start-up, one has to do everything that is required to grow the business. A marketer cannot depend on others, as he does in a large company to set up trade shows, arrange travel, and perform low-level marketing tasks. In a start-up, a marketer must take the initiative to make things happen. Workdays could start in the early morning hours and end late at night. There is no such a thing as a weekend. Therefore, think twice before moving to a start-up. A marketer can get great experience and become a fighter by going to a start-up, but he has to be ready to sacrifice his personal life for years. However, it may pay off in a big way.

At a start-up, a marketer can also get the experience needed to start his own venture in the future. Large companies do not teach the art of raising money and selling ideas formally. Large companies may give someone a comfortable life, but not always the satisfaction of knowing that an individual made the product a success. At a start-up, the success or failure of the product greatly depends on the ideas and efforts of a few people. A semiconductor start-up has to compete with fierce large competitors who are trying to destroy it. The team spirit of a start-up (the group united against an external enemy) is the competitive advantage a large company can hardly duplicate.

MORE TIPS FOR SUCCESS IN MARKETING

Semiconductor marketers have to sell their ideas first to their internal organization, then to the outside world. If a marketer can't sell his ideas, his analysis of a market segment, his passion about a product to engineers and financial people inside his company, then he cannot sell them to the outside world. Selling is a key component of leadership. Leading a group of people or a product requires that a marketer have passion about his product line and continuously think strategically, but act tactically.

A marketer must be able to create business opportunities by understanding the market dynamics and be the champion of his product line. This means a marketer ends up getting both the reward and the blame associated with the success or the failure of a product line. A good marketer takes ownership, not just in the development phase but in the selling phase of a product as well.

Marketing can be very challenging. Objectives and accomplishments are not always well defined and measurable. In addition, a marketer may work on a business venture or a product line for years before seeing the fruits of his labor. To make matters worse, the daily activities of a marketer are filled with interruptions, phone calls, and emails about problems that have to be resolved the same day or in short order. Marketers who cannot handle a constantly changing environment will become frustrated in their work. They are the ones who try to blame the system, people around them, or their company for their frustrations. These people have to once again examine if they want to be in the business side of semiconductor industry.

A successful marketer must develop the skills that prevent him from overreacting to different situations and news. He must believe in his ability to make things happen. Good marketers avoid too much anger and emotion. They keep a steady hand on the tiller to stay on course.

A good marketer learns how to balance various priorities better, and how to quickly move between short- and long-term business issues. He develops skills to be more articulate and assertive.

SUMMARY

There are those who believe good sales and marketing people are born with a gene to create and sell business opportunities. There are also those who believe marketing and sales can be taught to anyone. A mixture of the two might be the necessary ingredient for success. A marketer must make a habit of success. These include strengthening much needed critical skills, and development of a marketing mind-set to ensure the success of his business and his career.

CONCLUSION

There are huge opportunities ahead for the semiconductor industry; however, the challenges will be enormous. The preface of this book began with a review of the Internet bubble and the ensuing hangover in the industry. Perhaps the bubble bursting was a positive thing because it brought sobriety and thoughtfulness into our decision-making for some time to come. Given the extremely competitive era that our industry is entering, everyone in a company will need to execute as a team. We all need to be marketing-minded, rallying behind the vision of the company. Nevertheless, and most importantly, companies and marketing people must base their decisions on solid marketing principles to ensure their success.

This book provided an introduction to the field of semiconductor marketing by going through the steps it takes to bring a product to market as well as out to maximize revenues and profits once a product is introduced to a market.

For the benefit of the reader, we are providing a list of key points presented in this book that everyone should remember about the marketing field:

* There are three fundamental activities associated with marketing:

    1. Analyzing Markets
    2. Targeting Markets
    3. Finding the best way to sell products and/or services in the target market

* The main goal in semiconductor marketing is to maximize profits and minimize risks by leveraging sources of sustainable competitive advantage.

* The development of the roadmap is the primary tool for showcasing a semiconductor product strategy over a fixed timeline.

- Marketing is in charge of the initial product definition through an MRD and is the initiator for subsequent business justifications and gate reviews.

- Marketing must take an active (if not official) role in project management during the product development.

- A great deal of planning must go into product roll out and subsequent product promotion to ensure success in the marketplace

- Excellent technical support is a critical component to market success for semiconductor products.

- Forecasting is necessary to ensure commercial success of product lines especially legacy product lines.

- Optimal marketing organization depends on various factors such as the market segment and product strategy.

- A successful marketer develops a base of contacts and mentor upon whom he can rely to help solve difficult business problems and strategic questions.

- A successful marketer develops himself by obtaining relevant technical and financial knowledge as well as finding opportunities to develop critical skills necessary for the job.

- A successful marketer relies on knowledge as well as intuition, creativity and leadership to make good business decisions.

- All members of a company must adopt a marketing mentality necessary to ensure product and company success.

We hope that the information in this book will help the reader gain insight into the marketing process. We are convinced that all employees must understand and appreciate what it really takes to be successful in the semiconductor market. It is through this focus and knowledge that companies will grow strong in the post bubble era of extreme competition.

INDEX

A

B

C

GLOSSARY

A

ADC	Analog-to-Digital Converter
AMCC	Applied Micro Circuits Corporation
AMD	Advanced Micro Devices, Inc.
AOL	America Online, Inc.
APAC	Asia Pacific
ASIC	Application Specific Integrated Circuit
ASP	Average Sales Price
ASSP	Application Specific Standard Product
ATI	ATI Technologies Inc.

B

BOM	Bill of Materials
CDMA	Code Division Multiple Access

C

CE	Consumer Electronics
CEO	Chief Executive Officer
CES	Consumer Electronics Show
CMOS	Complementary Metal Oxide Semiconductor

D

DFT	Design For Test
DLP	Digital Light Processing
DSL	Digital Subscriber Line
DSP	Digital Signal Processing
DVD	Digital Video Disk

E

EDA	Electronic Design Automation

F

FAB	Fabrication Facility
FAE	Field Application Engineer
FPGA	Field-Programmable Gate Array

G

GPS   Global Positioning System

H

HP   Hewlett-Packard

I

I/O   Input Output

IC   Integrated Circuit

IDM   Integrated Device Manufacturer

IEEE   Institute of Electrical and Electronics Engineers

IP   Intellectual Property

IRR   Internal Rate of Return

ISV   Independent Software Vendor

IT   Information Technology

L

LAN   Local Area Network

M

MBA   Master's of Business Administration

MRD   Marketing Requirements Document

MSO   Multiple Services Operator

N

NPV   Net Present Value

O

ODM   Original Design Manufacturer

OEM   Original Equipment Manufacturer

OS   Operating System

P

PC   Personal Computer

PC OEM   Personal Computer Original Equipment Manufacturer

PCB   Printed Circuit Board

PDA	Personal Digital Assistant
PLD	Programmable Logic Device
PR	Public Relations

R

R&D	Research and Development
RAMDAC	Random Access Memory Digital-to-Analog Converter
RF	Radio Frequency
RMA	Return Material Authorization
ROI	Return on Investment

S

SAM	Serviceable Available Market
SoC	System on a Chip
SWOT	Strengths Weaknesses Opportunities and Threats

T

TAM	Total Available Market
TI	Texas Instruments
TSMC	Taiwan Semiconductor Manufacturing Corporation

U

| UMC | United Microelectronics Corporation |
| USB | Universal Serial Bus |

V

V.90	Modem connection protocol; allowing for speeds up to 56Kbps using standard (analog) phone line
VLSI	Very Large Scale Integration
VP	Vice President

W

| WHQL | Windows Hardware Quality Laboratory (Microsoft) |
| WINTEL | Windows-Intel |

ABOUT THE AUTHORS

AL SERVATI

is a co-author of the Intranet Bible (ISBN 1884133312), an early primer on setting up networked systems. Servati holds the position of Division Director of Conexant's Convergence Video business responsible for a team in charge of customer, product, and technical marketing. During his career at Conexant, Servati held several management positions including running Conexant's cable modem, consumer set-top box as well as IP-set top box businesses. Prior to joining Conexant, he served as co-founder and Vice President Marketing at Syntricity, the originator and leading provider of Web-native software and services for semiconductor yield improvement. Servati also held positions in marketing at Rockwell Semiconductor, and as a staff product engineer at Unisys Corporation, in San Diego, California. He holds a bachelor's and master's degree in electrical engineering, as well as an MBA from San Diego State University.

ANTHONY SIMON

is the author of The COT Planning Guide (ISBN 1931541981) a handbook for fabless semiconductor companies on IC design and manufacturing outsourcing strategies. Currently Anthony holds the position of General Manager for Pixelworks Corporation Japan based in Tokyo. Pixelworks is a semiconductor company focused on developing products for the digital television market. Prior to his position at Pixelworks, he held several marketing positions at Conexant Systems, including Strategic Marketing Director for Asia for Conexant's set top business division. His responsibilities included

235

overseeing business development, product strategy, and product promotion for the Asia region, especially focusing on market development
in China, Japan, and Korea. Prior to Conexant, Anthony served as a
product-marketing manager for VLSI Technology (acquired by Philips
Semiconductor) based in Rennes, France. Anthony holds a Ph.D. in political science from the University of Debrecen, Hungary, and a Master's
in economics from George Mason University, Virginia. He earned his
Bachelor's in electrical engineering from the University of Akron, Ohio.
Anthony is fluent in French, Hungarian and Japanese.

www.ingramcontent.com/pod-product-compliance
Lightning Source LLC
Chambersburg PA
CBHW022123050726

47590CB00002B/378